Crazy Chicken Lady 2 - In - 1

BECOME A MAGNIFICENT CRAZY CHICKEN LADY + 809 CREATIVE CHICKEN NAMES - THE BEGINNER'S 2 - IN - 1 BOOK FOR LEARNING TO RAISE AND NAME CHICKENS

LM TAYLOR

Contents

BECOME A MAGNIFICENT, CRAZY CHICKEN LADY

Introduction	3
1. FIRST STEPS TOWARDS YOUR CHICKENDOM	13
Things to Consider	13
Summary	21
2. WHICH COMES FIRST—THE CHICKEN OR THE EGG?	22
Breed Selection	23
The Chicken Life Cycle	32
How Old Should Your Chickens Be When You Buy Them?	36
Getting Your First Chickens	40
Hatching Eggs	42
Chick Care: Everything You Need to Know	48
When to Move Your Chicks to the Coop	55
Production Cycle of Layer Chickens	57
When Hens Stop Laying...	58
Summary	59
3. FLOCKS IN THE COOP	61
Chicken Coops 101	62
Building the Coop	65
Fencing	68
Predator-Proofing Your Chickens	71
Raising Free-Range Chickens	73
Roosters: The Big Question	76

Preparing for Storms and Severe Weather Events	78
Summary	81
4. HOW—AND WHAT—TO FEED YOUR FLOCK	**83**
How to Feed Chickens	84
What *Not* to Feed Chickens	92
Summary	94
5. TIPS FOR REACHING YOUR CHICKEN-RAISING GOALS	**95**
Tips for Raising Egg-Laying Chickens	95
Tips for Raising Meat Chickens	98
Tips for Raising Show Chickens	102
Your Chicken Chores	105
Summary	108
6. CHICKEN LANGUAGE	**109**
The Order of the Chicken	110
Managing the Pecking Order	113
Common Chicken Behaviors	115
Disciplining Your Chickens	119
Summary	123
7. WHAT'S WRONG WITH MY CHICKENS?	**125**
Common Health Problems	126
Diseases	128
Summary	133
8. CASH CHICKENS—NOT COWS	**135**
Making Money From Your Chickens	136
Summary	142
Afterword	143
Bibliography	147

809 CREATIVE CHICKEN NAMES

Introduction	157
1. TRADITIONAL	159
2. A LITTLE BIT COUNTRY...	162
Chicken songs:	162
3. A LITTLE BIT ROCK 'N' ROLL...	166
Chicken Songs:	166
4. THUG LIFE / HIP HOP / R&B	168
Chicken songs:	168
5. MORE FEMALE VOCAL POWERHOUSES	170
6. OVERLY AGGRESSIVE ROOSTERS	172
7. VIOLENT HENS	174
8. FOR THE FIGHTERS	176
9. SO WRONG...BUT SO RIGHT...	178
10. JUST BECAUSE I THINK THESE ARE FUNNY	180
11. MORE FOR ROOSTERS	183
12. ADULT THEME	185
13. DUOS, TRIOS, AND ENSEMBLES	187
14. INSPIRED BY OUR FAVORITE ADULT BEVERAGES	190
15. HISTORICAL	192
16. BASED ON GAMES	194
17. LITERARY GREAT AND INSPIRING ARTISTS	196
18. SPORTS GREATS	198
19. STRONG FEMALE PERSONALITIES	201
20. FEMALE LEADS	203
21. FASHION ICONS AND LUXURY CARS	205
Luxury Vehicles	206
22. FICTIONAL CHARACTERS	208
Bibliography	211

© Copyright 2023 - **All rights reserved.**

The content contained within this book may not be reproduced, duplicated or transmitted without direct written permission from the author or the publisher.

Under no circumstances will any blame or legal responsibility be held against the publisher, or author, for any damages, reparation, or monetary loss due to the information contained within this book, either directly or indirectly.

Legal Notice:

This book is copyright protected. It is only for personal use. You cannot amend, distribute, sell, use, quote or paraphrase any part, or the content within this book, without the consent of the author or publisher.

Disclaimer Notice:

Please note the information contained within this document is for educational and entertainment purposes only. All effort has been executed to present accurate, up to date, reliable, complete information. No warranties of any kind are declared or implied. Readers acknowledge that the author is not engaged in the rendering of legal, financial, medical or professional advice. The content within this book has been derived from various sources. Please consult a licensed professional before attempting any techniques outlined in this book.

By reading this document, the reader agrees that under no circumstances is the author responsible for any losses, direct or indirect, that are incurred as a result of the use of the information contained within this document, including, but not limited to, errors, omissions, or inaccuracies.

Become a Magnificent, Crazy Chicken Lady

MAKE A NEW CACKLING (LITERALLY) CIRCLE OF INSECT-MURDERING BFFS, LAY YOUR OWN EGGS (NOT LITERALLY), AND ENTERTAIN YOURSELF LEARNING TO RAISE CHICKENS

Introduction

Lord, help me be the person my chickens think I am.

— UNKNOWN

Roosters are notoriously early risers. It's barely light and they wake you with their crowing, celebrating the new day and encouraging you to enjoy the dawn with them. The wild birds aren't even singing, and it could still be dark out. But the roosters are insistent, so you stumble out of bed, grab a cup of coffee, and head out to your chicken coop. You're probably still in your pajamas, which are soon soaked with dew, but the chickens won't notice your bed hair or questionable fashion-sense.

The roosters relax as you approach, and you can hear the hens' soft clucking as they go about their business—it's an oddly comforting sound. As the sun peeps over the horizon, you

INTRODUCTION

collect your first eggs of the day from the nest boxes; they are warm and brown. Later, you will have them for breakfast. Or perhaps bake a sweet treat for the family using your very own eggs. The yolks will be sunshine yellow and nutritious, way tastier than those from the store.

Keeping chickens is extremely satisfying—and you'll be surprised at how quickly these birds will nest in your heart. You'll soon be referring to them as "my girls," you'll know all their individual characters and quirky habits and assign names accordingly, and they'll be as much a part of your family as your children are.

As a backyard chicken keeper, you'll join numerous people who took up the hobby during the Covid-19 pandemic—and are still going strong. The American Pet Produce Association says that while 8% of Americans owned chickens in 2018, that jumped to 13% during 2020 as a result of the pandemic (Caley, 2021). Backyard chicken keeping has exploded, with millennials and Gen Z leading the charge due to concerns about supply chains and food security.

After experiencing empty food shelves during the pandemic, people started becoming more self-sufficient in food provision. Many people also want to know where their food comes from—and what better way than to raise it yourself? But there's more to keeping chickens than food security. People raise chickens to enjoy fresh farm eggs and healthy meat or to have a hobby they can share with their children—although many fall in love with their chickens once they start and just can't bear to let them go. There's official support for small-scale chicken keeping too. As of 2015, 93% of America's most populous cities allowed residents to raise backyard chickens ("How Many People," 2022).

THE FASCINATING HISTORY OF BACKYARD CHICKENS

In the early 1900s, chickens were regarded as unimportant livestock. The meat was considered a delicacy and eggs a luxury. During World War I, France ran out of fresh food and the United States War Relief Services sponsored chicken farming there. In 1918, after the U.S. entered the war, the government campaigned for people to start keeping backyard chickens to provide food (Harm, 2019).

After the U.S. entered World War II, Americans were again urged to start backyard gardening and chicken raising. Following post-war food shortages, people were encouraged to plant victory gardens and keep chickens. By 1943, there were 18 million victory gardens countrywide, mainly in cities (Harm, 2019).

Victory gardens dwindled when food production methods changed. Chickens were raised on large farms and eggs became so cheap that many farmers went out of business. Supermarkets began and people swapped rural life for suburban convenience. Chicken farming practices also became a cause for concern.

Things came full circle in the 1990s when people endorsed sustainable living, organic farming, and backyard chicken keeping. The rest, as they say, is history.

INTRODUCTION

BENEFITS OF RAISING BACKYARD CHICKENS

Big Rewards for Little Effort

Everybody loves chickens—they are largely self-sufficient and don't require special care. Once you've set up your coops and runs, the chickens only need to be fed and watered. You will need to clean the coop and watch for diseases, but that's about it. Chickens like entertainment, so install a coop swing or several perches. They need to get sufficient exercise; make sure the run is large enough to prevent arguments. Compared to keeping other livestock, chicken keeping is a breeze.

Fresh Eggs by the Dozen

The main reason why people keep chickens is—you guessed it—for the eggs. Whether you like sunny side up, hard-boiled, scrambled, or making egg-based dishes like quiches or souffles, chickens will provide this bounty in spades; nearly every chicken keeper says, "I get *so* many eggs." If you have free-range chickens, they'll have less cholesterol, more vitamins A and E, beta-carotene, and healthy omega-3 fats. They're much tastier than store-bought eggs because they're harvested fresh and eaten soon thereafter.

More Than Eggs

Chickens don't only provide eggs. They're a source of organic meat—and if you've raised them yourself, you know that they've been fed only the best, additive-free foods.

Their feathers can also be used to make your own down pillows, cushions, crafts, and more.

Combating Food Waste

Keeping chickens will reduce or even eliminate food waste, as chickens can eat leftover human foods as treats. However, table scraps shouldn't comprise more than 10% of their diet—chickens aren't garbage cans (Schneider, 2018).

Gardening Buddies

Chickens are excellent natural pest controllers. Besides weeds, grass, and other plants, chickens forage on numerous protein-rich insects. They also provide valuable manure if you're growing fresh produce or ornamental plants. Of the commonly used organic manure including cow, pig, and horse, chicken manure has the highest concentration of nitrogen and phosphorus.

Entertainment

When you have chickens, you'll have your own movie channel in your backyard. Chicken owners spend hours watching the antics of their flocks.

One chicken keeper I know has a hen that flies up and perches on her owner's shoulder whenever she uses her riding lawn mower. The hen only does this when she mows the lawn, for reasons no one understands. I guess she just likes to cruise and feel the wind in her feathers.

Another chicken keeper said her hen always used to lay her daily egg on a particular chair on the porch. One day, something frightened the hen just as she was preparing to lay. The chicken ran into the house, squawking and flapping, and laid her egg as she was running.

INTRODUCTION

Friendly Pets

Some chickens are, first and foremost, family pets. Chickens have fierce, determined personalities, but if you choose a gentler breed, they can easily become part of your family, endearing themselves to everyone.

Children learn about nature, and themselves, when they help raise backyard chickens. Not only do they discover more about chicken biology, but they also develop a better understanding of where their food comes from and how to care for other living creatures.

Sustainable Living

If you're keen to start homesteading, chickens are a great first step. Your carbon footprint will be reduced and much of your kitchen waste can be recycled. There is a growing demand for pasture-raised, organic chickens and eggs. You'll likely have more eggs than you can possibly eat, so you could sell your excess eggs and organic chickens to recoup your costs.

Health and Lifestyle Benefits

Besides the obvious benefit of eating fresh eggs and meat, chicken keeping is a physical activity and will help keep you fit and active. Caring for living creatures will lift your mood and reduce stress.

CHICKENS AS GARDEN FAIRIES

If you have a garden, chickens will make cultivation and gardening easier, especially if you are growing vegetables. You might need to share your produce with them, as chickens enjoy fresh food just as much as you do, but the benefits far outweigh the small sacrifice. Cultivate a little extra to allow for the chickens' activities.

No-Till Naturally

Chickens scratch the soil, looking for juicy insects, worms, seeds, and grit. Their long toes and sharp claws disturb the soil, aerating it and helping to break down organic matter. Put your chickens to work if you want to break ground for a new crop or till the soil without much disturbance.

Their natural scratching activities efficiently dig in manure and compost, improving the soil. If you're concerned about chickens damaging your garden plants, rotate the birds through different parts of your yard and fence off any no-go areas.

Free Compost

Chickens will fertilize your garden with their manure. Chicken manure makes fantastic fertilizer—it contains nitrogen, phosphorus, potassium, calcium, and other nutrients and trace elements that make plants thrive. This is wholesome, organic, and saves you time and money, as you won't need to buy as many additives for your garden soil. Eggshells are excellent for your compost heap, as they are rich in calcium. Bake them in the oven so they become brittle before crushing them and adding them to your compost.

Healthier Chickens and Eggs

Raising chickens yourself means you're getting the healthiest, most nutritious eggs. The eggs taste better and are of better

INTRODUCTION

quality than store-bought ones. Feeding your chickens excess produce, especially if you grow your own, can improve the flavor and color of their eggs.

You can grow several herbs for chicken forage that will keep them healthy. Chicken-friendly herbs include dill, cilantro, tarragon, and thyme, which contain antioxidants. Protein-rich herbs include fennel, basil, coriander, and mint.

Reducing Feed Bills

Feeding chickens high-quality, store-bought feeds can be expensive. If you have a large lot, grow herbs and plants like sorghum, corn, barley, and squash for both yourself and your birds to reduce feed costs. If you have a worm bin, feed ground eggshells to your worms and feed the worms to your chickens. Alternatively, use the vermicompost the worms generate to improve your soil, so you can grow more chicken food.

SUMMARY

More people everywhere are keeping backyard chickens and discovering the benefits of having these feathered friends. Chickens are one of the best small livestock types to raise. They are self-sufficient and reward you with tasty, organic free-range eggs and meat that are much tastier and healthier than those in stores. Chickens improve your garden by naturally tilling the soil, providing free manure, recycling food or fresh produce waste, and controlling pests and weeds. They can teach children about raising livestock and caring for living creatures. Chickens are also hilarious in their own right, providing hours of free entertainment by just being themselves. By selling chicken products like free-range eggs, organic meat, and feathered items, you can generate extra income.

In the next chapter, you'll discover important considerations to bear in mind before you start investing in chickens. Do you have the space, time, and energy to raise chickens? Do you have the cash for the start-up costs? And what will you do when your hens stop laying, or chickens fall ill?

CHAPTER 1
First Steps Towards Your Chickendom

People who count their chickens before they hatch act very wisely, because chickens run about so absurdly that it's impossible to count them accurately.

— OSCAR WILDE

Keeping chickens is a fantastic hobby with numerous benefits. However, while this can be a very rewarding undertaking, it's not all roses. You need to consider several aspects of raising these productive birds before you start.

THINGS TO CONSIDER

So, you're all fired up and ready to join your local backyard chicken community. But before you do, there are a few things you need to check on.

Is It Legal in Your Area?

It may sound strange, considering how many people are joining the backyard chicken craze, but your local laws might prohibit you from keeping small livestock if you live in town. Raising chickens in urban and suburban areas is not regulated by Right to Farm laws, which protect farmers and homesteaders from nuisance lawsuits (MacLean, n.d.). Every town has its own chicken-keeping rules, and these might even vary in different parts of a particular town or city. Unfortunately, there are no central databases, so you'll need to gather the information the old-fashioned way.

1. Contact your local government officials to learn about any chicken-keeping ordinances. Begin with your planning board, county clerk, or animal control department.
2. But even if your local municipality *does* allow backyard chicken keeping, there may be restrictions or regulations. These might pertain to flock size, coop building, or space required per bird. You might be allowed to raise hens but not keep roosters. Ask if permits are required, whether you need to inform your neighbors or obtain their permission, or if your Homeowners Association will allow you to keep chickens. Ensure that you understand all the rules and regulations before embarking on your chicken venture.
3. Get a copy of the local chicken ordinance and keep it on file—use this to address any concerns your neighbors may have over time.

4. If you are not allowed to keep chickens or build coops, campaign to have the laws changed so that you, and others, can keep backyard chickens to ensure sustainability and food security.
5. Talk to the neighbors. While your local officials might be happy for you to raise chickens, your neighbors might have doubts. Consider how your activities could affect them. Once you are up and running, invite them to meet your girls. Sharing your eggs with your neighbors can't hurt either.
6. When you design your coop, factor in your local conditions and your birds' requirements.

Do You Have Sufficient Space?

While you don't necessarily need a large amount of space to keep chickens, the question of how much room they need is surprisingly complex, as there are many variables. The amount of space required varies according to the breed, the birds' age, the proposed size of your flock, and whether they will have outdoor access. Chicken coops must have sufficient air circulation without becoming stuffy in summer and drafty in winter. Your chickens will need space to avoid one another so they don't become irritable or aggressive. There can also be seasonal variations to take into account. For example, if you live in an area with very cold winters, your chickens might need to be cooped up for long periods.

If you have a small breed, like Bantams, you can get away with as little as two square feet per chicken if they are free ranging. A medium-sized breed, like Leghorns, will need at least three square feet per chicken, while Plymouth or other larger breeds need at least four square feet. So, if you have 10 chickens and are considering the minimum space requirements for a medium-sized breed, you would be looking at a coop size of at least thirty square feet for free-ranging chickens (Guest Author, 2020).

Allow for more space rather than less when planning your coop—chickens can get cabin fever if confined in too small a space. This can trigger aggression and stress, which means less egg-laying and tougher meat.

Recommendations for Urban Poultry

- Laying hens should have a minimum of 3–4 square feet per hen indoors and 10 square feet per hen of outdoor space (the run) ("Space Allowance," n.d.). An outdoor run should be at least this size for all breeds and larger if possible.
- A large run may not be practical in an urban backyard, so you might need to opt for a smaller one, which would become more of an exercise yard. It needs to be managed appropriately, as the chickens will remove all the plant material. Ensure that it is in a well-drained location so it doesn't become muddy and wet when it rains.
- Meat chickens also need at least 3–4 square feet of indoor space per bird ("Space Allowance," n.d.). Broiler chickens grow fast, so they're unlikely to need an outdoor run, but you can provide one if you are keeping slower-maturing breeds.

How Big Should Your Flock Be?

The short answer is: from my own experience, start with a minimum of four.

Why not one? Well, one is, of course, the loneliest number. Chickens are social, flock-oriented animals. They are not happy on their own.

Why not two? If you start with two chicks, and one dies, one is once again the loneliest number.

Why not three? Actually, three might be okay, but it's a BARE minimum in my opinion. But even with three, your chances of ending up with one single chick are greater than you'd think. Also, if you get more chickens in the future, it is easier - and safer - to introduce a group of chickens to new newcomers than just your lone (and probably sad or potentially paranoid) bird.

While we are on the topic of starting small and adding more chickens later, I'll say that this is a logical, and common thought process for new chicken ladies. We think "I'll try this whole Crazy Chicken Lady thing out on a small scale to start, and I'll get more later once I know what the hell I'm doing...". Responsible planning! That being said, when I got my first batch of guinea hen chicks, I got 18. But I have 300 acres so I could do that. My logic was, caring for four birds seems like the same amount of work as caring for 18, so why not? Also, I am not a bird-cuddler. I am not concerned with befriending my birds on the level that many other Crazy Chicken Ladies do – so getting 18 was no big deal to me. If you envision being BFFs with your chickens, having them curl up in your lap, ride on your shoulder while you cruise on the riding lawn mower, or whatever else chicken-cuddlers do, you might want to keep your numbers lower to start with so the chicken-bonding can commence.

Okay now – at this juncture in your chicken education, I absolutely must familiarize you with "Chicken Math". Let's say you've decided to start small with a few chickens and then add to your flock later. That's the concept. The reality is that you decide to get X number of chickens for starters, but what actually happens is you end up getting at least three times that number—or more.

Because chickens are self-sufficient, not to mention adorable—and there are so many breeds, colors, and egg differences—it isn't long before you start adding rapidly to your flock. You might pass a feed store or see an advertisement for day-old chicks and think, "Maybe I can have a few more." If you hang out on online communities for backyard chicken keepers, you'll see lots of examples of people succumbing to the phenomenon. In extreme cases, backyard chicken keepers even start branching out to keep ducks, geese, and maybe even a few goats or a cow.

Before you succumb to chicken math, be realistic with yourself about how many you are going to have, and plan accordingly. I know they're adorable. I know you want to get 267 different colors/breeds. They are like a living, strutting, clucking version of the old china figurine collections all of our great-grandmothers had. That's completely and utterly awesome – you do you, girl. Just please, please, PLEASE ensure that you treat your gals (or maybe gals and guys) humanly and give them the appropriate amount of space. Bear in mind your time, finances, and other resources. Whatever you decide, be certain that you are able to give your birds the very best care.

Costs of Raising Chickens

This depends to some degree on where you live, but you should save up for start-up costs like your chicken coop, lighting, and food and water containers. If you are raising chickens for meat, you might need an egg incubator, which can hatch several birds at once. Incubators effectively take the place of hens sitting on the eggs to hatch them.

You could buy pullets—teenage chickens—rather than chicks, as these are almost ready to begin laying, and you won't need to care for them for the first eight weeks or so. This can save you money while bringing in income sooner to help cover costs (Taylor, 2023).

It is essential to use a reputable breeder when you buy chickens. You will get healthier chickens through selective breeding, so they're less likely to succumb to disease. The main breeds and costs of pullets are

Ancona: $3–$5

Easter Egger: $2–$3.50

Buff Laced: $4–$6

Silver Phoenix: $4–$6 (Cosgrove, 2021)

You'll need a coop to keep your birds warm and protect them from predators. The costs vary tremendously, depending on whether you are able to DIY from salvaged materials, build from scratch, or decide to buy a ready-made one. You could spend anything from $200–$700 on your coop (Cosgrove, 2021).

High-quality feed is essential to ensure your flock gets the right nutrients so they stay healthy and productive. You should budget $50–$60 per 40 lb bag (Cosgrove, 2021). To give you an idea of how long that will last, a laying hen eats about 1/4 lb of feed a day, equating to 1.5 lbs of feed a week. A 40 lb bag then feeds your chicken for 26 weeks. If you have 4 hens in your flock eating 6 lbs of feed a week, a 40 lb bag will last you about 6 weeks (Balam, n.d.).

You'll need a waterer or fountain to keep your birds cool and adequately hydrated. These cost between $25 and $50. Soft bedding will keep them comfortable and also warm in winter. Budget around $25–$50 for bedding (Cosgrove, 2021).

Depending on your situation, you might need to budget for a heater and lighting if you are in a cold climate, as well as fencing.

Your chickens will generally cost you around $200–$400 per year for a small flock kept in a coop (Cosgrove, 2021). The initial cost will be higher because you need to buy the coop, equipment, and other items. You might lose birds initially due to inexperience, but apart from that, you can raise chickens relatively cheaply.

Note: Prices are updated and correct as of January 2023.

Chicken Health

While they require no regular veterinary checkups, it's recommended you get your chickens vaccinated against Marek's disease, fowl pox, Newcastle disease, and several other ailments that affect chickens. Get your birds vaccinated rather than waiting for disease to break out in your flock, which may well end in tears.

SUMMARY

There are several things to check before you start keeping chickens. While they are relatively self-sufficient and inexpensive to keep, especially if you're selling their eggs or meat to help cover costs, there are other things to bear in mind. Establish whether your local municipality will allow you to keep chickens in your area, ensure that your space is large enough for chicken keeping, decide on the number of birds, and consider whether you can afford the initial outlay.

CHAPTER 2

Which Comes First—The Chicken or the Egg?

The key to everything is patience. You get the chicken by hatching the egg, not smashing it.

— ARNOLD. H. GLASOW

So, you've established that you can keep chickens in your local municipal area, you have the time required to spend on them, and the cash for the initial outlay and their care thereafter. Now you need to decide what breed you want to raise. Once again, there are several factors to consider before you begin, including your local climate, what you want to use the birds for, and so on. You could start your flock from eggs, buy day-old chicks, or buy older chickens—the choice is yours. Make sure that your choices are the right ones for you and your particular circumstances.

BREED SELECTION

Factors to Consider When Choosing a Breed

Choosing a breed is probably the most important decision you'll make as a backyard chicken keeper—apart from deciding to raise poultry in the first place.

When choosing a breed:

- Consider your local climate. Certain breeds adapt to fluctuating temperatures better than others. Some chickens do better in hot climates, while others prefer cooler climes, so choose a breed that suits your region. Andalusians and Leghorns do well in warm areas, while Australorps, Barred Rocks, and Buff Orpingtons do better in cool places, for instance.
- Why do you want to keep chickens? Do you want to raise them for eggs, meat, or both? If it's eggs you want, do you prefer a particular egg size, or do you want a certain color? All chickens lay eggs, but some are more prolific. Medium layers are sufficient for a family, but some breeds produce more. A good layer will produce 5–6 eggs a week in spring and summer (Steele, 2023).
- Breeds used for meat mature faster than those preferred for egg laying and are poor layers, so you're unlikely to get the best of both worlds.
- Do you want to raise birds for show? Poultry shows and exhibitions are becoming popular.
- The space you have available for raising chickens will determine what breeds you select. Some chickens

don't mind being cooped up, while others prefer free-range situations.
- Consider the temperament of different chicken breeds—and what would best suit your family. If you intend to have a very small flock, chances are the birds will be more like pets. If you want to involve your children, find a breed that will make the experience enjoyable. Roosters are aggressive and you may wish to avoid them if chicken keeping will be a family affair. Docile breeds suitable for children include Cochins and Silkies. Bantams are smaller and might be less intimidating.

Best Breeds for Egg Production

Here are some of the best breeds for eggs:

- Plymouth Rocks lay around 200 light-brown eggs annually—perfect if you want to collect eggs nearly every other day (Morning Chores Staff, 2018). They are friendly, relatively docile, and easy to raise. These chickens have several varieties and their eggs are usually large. They are broody hens—chickens that lay and incubate fertile eggs—and make good mothers. They don't mind the cold and can be used for meat production too.
- Barnevelders also lay in the region of 200 light-brown, small to medium-sized eggs a year (Morning Chores Staff, 2018; Ohio State University, 2017). These chickens don't fly high, so they are great for

backyards. They are docile, calm, and will adapt to both being kept in a coop or raised free-range.
- Anconas are also in the 200-eggs-a-year league, producing large, white eggs (Morning Chores Staff, 2018). They like to fly, so they need an environment where they are able to do this freely. They are active and tend to be wild, so they do best in free-range situations. Varieties include single comb and rose comb.
- Hybrid chickens are excellent layers, averaging around 280 eggs a year—and they don't require a lot of food either. The eggs are brown and medium-sized. If you choose the Golden Comet variety, they won't become broody and are easier to keep (Morning Chores Staff, 2018).
- Buff Orpingtons lay only 180 eggs a year by comparison because the hens are inclined to become broody (Morning Chores Staff, 2018). They can become very tame, however, even eating from your hands.
- Leghorns are white chickens laying around 250 white eggs a year (Morning Chores Staff, 2018). Eggs are anything from medium to extra large. There are several varieties, and if you spend time with them, they will delight your family and yard.
- Hamburgs lay 200 very white eggs annually, the size depending on what you feed them, as well as their environment (Morning Chores Staff, 2018). They prefer larger spaces and love foraging, so ensure your yard is big enough. There are several varieties, all with eye-catching plumage consisting of black, white, and

golden feathers (Morning Chores Staff, 2018; Ohio State University, 2017).

Breeds for Different Egg Colors

Most chicken eggs start off as white. Several environmental factors influence egg color. Excess sun exposure or lack of adequate drinking water can make the eggs lighter. If the hen has red earlobes, she is more likely to produce eggs in shades of brown—as well as green, blue, or even pink.

White as Snow

If you want white eggs, consider adding the following breeds to your flock:

- Sebrights don't lay well and are difficult to breed, but they produce small, white eggs.
- Minorcas lay large, white eggs all year and are ideal for backyard chicken keepers, as they love people. They are a flight risk, though.
- Silkies lay tinted eggs. They are gentle and make great mothers.
- Hamburgs don't eat as much as other breeds and lay around four eggs a week (Morning Chores Staff, 2018).
- Andelusians are decent layers, providing about three eggs a week and fifty eggs a year (Morning Chores Staff, 2018).

Brown Is Beautiful

Brown eggs come in several different hues, ranging from beige to dark brown. If you want to have brown eggs, add the following to your flock:

- Brahmas are gentle birds and good layers, giving you around three eggs per week (Morning Chores Staff, 2018).
- Cochins produce enormous eggs in large quantities but love their food. Be careful that they don't become fat, or they will stop laying.
- Chanteclers are friendly birds but tend to become broody. If you want plenty of chicks, this is the breed to choose, as the mothers lay large clutches and look after them well.
- Delawares are the perfect choice for jumbo eggs. Expect to get around four eggs per week (Morning Chores Staff, 2018). They don't need as much care as some other breeds and are less likely to get broody.
- New Hampshire Reds are perfect for backyards as they're docile and calm. They lay about 200 eggs a year (Morning Chores Staff, 2018).
- If you prefer darker brown eggs, try Barnevelders, Cuckoo Marans, and Blue Copper Marans.

In the Pink

You've heard of pink gin and pink champagne. Now you can have pink eggs too! There are a few breeds that lay beautiful pinkish eggs.

- Australorps are excellent layers, producing 200–240 eggs a year, ranging from light brown to pink (Morning Chores Staff, 2018). Check with the seller that the chickens actually do lay pink eggs.
- Light Sussex are adaptable, friendly souls that are easy to raise. Their eggs vary from light brown to pink depending on how much sun they receive.

No More Blue Days

Blue Mondays will be a thing of the past with these special hens who lay amazing blue eggs.

- Cream Legbars are best raised as free-range chickens. Their eggs come in different shades of blue, and unusually you can distinguish between male and female birds on hatching.
- Araucanas, a very rare breed, are unique in that they are black with no tail head. Their eggs are always blue.

A Meaty Decision

If you want healthy, organic meat, then raising your own poultry is perfect because you know what they are eating. Chickens raised for meat are called broilers and grow much faster than layers. They may reach 10 pounds in 10 weeks—big enough to feed a family (Morning Chores Staff, 2018).

So, what's the best broiler breed?

- Jersey Giants can weigh as much as 13 pounds when they're ready for slaughter and resemble turkeys because they have so much meat on them. They're black, blue, and white and reach maturity in less than a month (Morning Chores Staff, 2018).
- Freedom Rangers reach maturity later than other broiler breeds (9–11 weeks) and eat insects as well as corn-based feed (Morning Chores Staff, 2018). The delicious flavor of the meat is well worth waiting for. They also eat low-protein foods but need space.
- Cornish Crosses are meaty chickens favored by both commercial and backyard chicken keepers. They do require more feed, however.
- Bresse is a much smaller breed and accordingly produces less meat, although it tastes very good. These chickens take longer to mature than other meat-producing breeds and are therefore more expensive to raise.

Dual-Purpose Breeds

If you want to keep chickens for both eggs and meat, then there are some dual-purpose breeds that will enable you to do this. Besides Maran and Sussex, which were discussed previously, you could buy one of these:

- Potchefstroom Koekoeks are a mixture of breeds that have been crossbred to ensure that the best features of each dominate. They are usually black and white.
- Wyandotte are popular show birds available in various colors. They produce most of their eggs in

fall and winter (Morning Chores Staff, 2018). They are peaceful birds, ideal for backyard situations.
- The Turken looks oddly like a turkey. Originally bred in Germany, they are calm, friendly, and easy to tame.

Chickens for Cold Climates

Certain breeds can withstand very cold temperatures, even going down as low as 10–32 °F (Hudson, 2022). Overnight temperatures in the coop are often around 30–40°F in winter, as each chicken generates sufficient heat to keep everyone warm. Chickens have a high metabolism, which helps them withstand cold. Their bodies' resting temperature ranges between 105°F and 109°F, with their heartbeat averaging 400 beats per minute ("How Cold is Too Cold," 2021).

Chickens able to withstand very cold temperatures have certain adaptations, like smaller combs (larger ones could get frostbitten), more feathers for extra insulation, and featherless legs (feathered legs could get covered with snow and slush, leading to frostbite.)

Cold-adapted breeds include:

- Ameraucanas, which have a beard and pea comb, are the preferred option for cold climates and lay coveted blue eggs. Laying declines markedly after the first year, however (Morning Chores Staff, 2021).
- Buckeyes, which are friendly and peaceable, are an excellent choice for backyard chicken keeping. They have small combs and feathery plumage.

- Rhode Island Reds, which have copious feathers, are ideal for cold climates but they can tolerate warm weather too.
- Australorps, which have thick, heavy feathers, are ideal for withstanding cold temperatures.

Breeds for Hot Climates

There are chicken breeds that do better in warmer regions, where temperatures regularly exceed 90 °F (Morning Chores Staff, 2021). These chickens have fewer feathers, larger combs, and lighter colors, all of which help to prevent heat exhaustion. Smaller chickens also thrive in the heat.

Breeds suitable for warmer climes include

- Fayoumi, which have striking plumage, are sought-after for poultry exhibitions, although they require a lot of space. If you have experience raising backyard chickens, this might be an option.
- Leghorns, which are perfect for hobby chicken keepers and lay eggs when they are young.
- Golden Buff chickens, which are ideal for both warm and cool regions. The females are small, relaxed, and easy to breed.
- Barred Plymouth Rocks are suitable for both warm and cold climates.

Bantam Chickens

Bantams can be miniature versions of other breeds and also have their own unique breeds. They are a quarter of the size of regular chickens and are ideal for backyards (Morning Chores Staff, 2018). They eat less and make good pets.

Most are excellent layers, normally producing 4–5 eggs per week, although these are smaller than eggs from larger chickens (Morning Chores Staff, 2018). Bantams are sociable, calm birds that make great children's pets and are very entertaining. The hens are good mothers and remain on their eggs when brooding.

Bantams don't require as much space as regular chickens and are perfect for urban and suburban backyards; you could keep more Bantams in the same space. They need smaller chicken coops and less food and water, so they are suitable for chicken keepers on a budget. Some Bantam breeds are striking enough to be used as show chickens.

Bantams can be raised with other breeds, but make sure the bigger birds don't bully them. You also don't want them mating with a regular-sized rooster! If you decide to raise Bantams and other breeds together, set up separate feeding stations to accommodate the different feed requirements. Bantams require high fences and secure boundaries as they fly well. Make sure your fencing is in good shape so they don't squeeze out of any gaps.

THE CHICKEN LIFE CYCLE

To help you make informed decisions about your chickens, it's a good idea to have some idea of their life cycle—how they start off in life (yes, the egg comes first), how long they're productive for, and their longevity.

Rather like humans, roosters court their chosen females to win their favor, partly by feeding them food morsels (tidbitting). It's not easy, though. Hens are picky. They want a handsome male that can find good food sources and protect the flock. If he's the strongest and healthiest roster, assuming there are more in the flock, she'll choose him.

Hens lay an egg every 25–27 hours ("Beginner's Guide," 2018). These aren't fertilized unless the rooster mates with her. Mating is quick—less than a minute—with the rooster mounting the hen's back. Hens can keep sperm inside their bodies for as long as three weeks and are even able to eject it if they decide they don't like the rooster after all ("Beginner's Guide," 2018). Fertilization occurs in the infundibulum—the reproductive tract that leads to the ovary. The egg is only there for about 15 minutes, so there's a very small window for fertilization (Lesley, 2020).

The hen continues to lay fertile eggs, gathering them in the nest until she feels she has a sufficient amount. She now becomes a "broody hen," diligently incubating the eggs for 21 days ("Beginner's Guide," 2018).

The Embryo

After fertilization, the embryo begins developing—in as little as two days, there is already tissue forming and blood circulation (Lesley, 2020). By day 14, the embryo starts getting ready to hatch, its head moving into the pipping position (Lesley, 2020). The following day, the embryo is fully feathered and consumes the egg white. On day 19, the yolk sac enters the embryo's body and the bird occupies all the space inside the egg except for an air cell (Lesley, 2020).

The next day, the yolk is absorbed, and the embryo begins to breathe air and becomes a chick. The yolk sustains the chick for up to 72 hours after hatching ("Beginner's Guide," 2018). This is when the pipping starts. On day 21, the chick usually hatches (Lesley, 2020).

The Chick

The hen attends to the chicks' needs. They remain under her wings for the first few days, keeping warm. Move the mother hen and her chicks to a safe, secluded place away from the flock during this time. Provide some starter crumbs for the chicks, together with room-temperature water to which electrolytes have been added. This will get everyone off to a good start. You will need to change the water several times a day as chicks are messy.

Chicks sprout their first real feathers about two weeks after hatching and molt for the first time at four weeks ("Beginner's Guide," 2018; Lesley, 2020). They grow rapidly in the next 3–4 weeks. Some breeds, like Cornish Crosses, grow so fast that they weigh 8–10 pounds when they are just 8 weeks old. By the fifth week, the chicks are able to control their own body temperatures ("Beginner's Guide," 2018). The next molt happens at 6–12 weeks, and this is when the males' feathers will differ from the females' (Lesley, 2020).

Chicks learn fast, especially if a broody hen is raising them. She teaches them what foods to eat, what to avoid, and where danger lurks. Watch them as much as you like but interfere with the chicks at your peril. A mother hen is very protective and might attack you.

If you are raising the chicks yourself, they can go outside at eight weeks but ensure they are protected from predators, hot sun, wind, and drafts ("Beginner's Guide," 20218; Lesley, 2020). Take them inside immediately if they appear stressed. Start introducing new foods like mealworms, grains, greens, and so on.

Pullets and Cockerels

Chickens become "teenagers" at 4 weeks old, and this life phase continues until they are 12 weeks (Lesley, 2020). At this stage, the females are known as pullets, while the males are called cockerels. At this age, the chickens are awkward and gawky, just like human teens. You'll be able to tell the males from the females, so separate the sexes now. The pullets start developing their pecking order—the hierarchical system regulating chicken society, which will be discussed in Chapter 6.

Once the pullets and cockerels are about two-thirds of the size of adults, introduce them to the flock (Lesley, 2020). Do this gradually, in a large space with nooks and crannies where the younger birds can hide if necessary. The young chickens now have to fit into the flock's existing pecking order. This can look traumatic but don't interfere unless birds are being wounded or injured. This is a natural process.

Hens and Roosters (Adults)

There is a little controversy as to when a pullet becomes a hen, with some saying this occurs at one year of age and others when she lays her first egg (Lesley, 2020). Pullets become hens and cockerels become roosters when they are sexually mature.

Hens lay their first eggs when they are about 20 weeks old (Lesley, 2020). These might be small and oddly shaped at first, but this is normal. If you have older hens, the younger ones will learn to use the egg boxes. If not, you will need to show them what to do.

HOW OLD SHOULD YOUR CHICKENS BE WHEN YOU BUY THEM?

Another decision you will need to make if you intend to raise backyard chickens is how old they should be when you purchase them. Options include buying fertilized eggs, day-old chicks, pullets, point-of-lay birds, or adult birds. You could even get rescue hens. Your chicken-keeping goals will influence the age of the chickens you choose.

If you want pet chickens or to get your children involved in raising them, then fertilized eggs or day-old chicks are great options. It's the cheapest way to get started, although you will need to buy special equipment. You'll require an incubator if you go the fertilized eggs route and a brooder (heated enclosure used to raise baby birds) to keep day-old chicks warm. Remember that it's almost impossible to determine whether you have hens or roosters when you buy fertilized eggs or day-old chicks.

For people who want egg-laying chickens and would like to get started quickly, pullets are a good idea, as they aren't quite old enough to produce eggs but won't need a brooder or heat. Point-of-lay birds are another excellent choice as they are about to start producing eggs. You can purchase adult birds that are proven layers, which might be more reliable.

Other factors influencing your choice include how much money you want to spend, how much chicken-keeping experience you have, and how much time and effort you can give the endeavor.

Be aware of the advantages and disadvantages associated with each option.

Pros and Cons of Fertilized Eggs

Fertilized eggs are a wonderful, natural way to obtain chicks—you can either use an incubator or a broody hen if you already have chickens. Watching chicks hatch can be an exciting experience for children, and they will bond with the chicks. This is one of the cheapest options for getting chickens (sometimes eggs are given away for free). Hand-raised chickens make excellent pets. You might obtain rare breeds affordably or have access to a wider range of breeds with specific genetic traits. If a breeder is far away, obtaining fertilized eggs will make it easier to get the chickens you want.

However, fertility and hatching rates are very variable. You might have to cull roosters you don't want or can't keep. You will need special equipment—an incubator and a brooder. While you can use a broody hen to hatch your eggs, success isn't guaranteed.

Pros and Cons of Day-Old Chicks

The sheer cuteness of day-old chicks makes them irresistible—and raising them is another amazing opportunity for your kids. Breeders often vaccinate and may even sex them for you, so you won't need to worry about roosters if you find a breeder that provides these services. You'll also get all the benefits of raising chicks without needing an incubator.

Raising day-old chicks can be hard and is often time-consuming. You will need a brooder so the chicks stay warm and healthy. They are delicate and can succumb suddenly to diseases or ailments. If you want eggs, the chicks will only be old enough to lay after about 4–5 months (Smith, 2020). Day-old chicks are usually only available in common, popular breeds, and you could get roosters. If you already have chickens, you might need to separate the chicks from the adults for a while.

Pros and Cons of Pullets

Pullets are more economical than fertilized eggs or day-old chicks, as they don't require special equipment, even though they cost more. The birds are old enough to sex, so you won't get roosters you don't want. They are hardy, adaptable, often make excellent pets, and are usually vaccinated by breeders.

These birds can be hard to find and are frequently only available in common breeds. If you already have chickens, you need to be cautious about introducing them, as older, larger chickens might bully them. If you want eggs, you will need to wait a month or two before they start laying (Rachael, 2020).

Pros and Cons of Point-of-Lay Birds

These are a more expensive option, but the big advantage is that the birds are beginning to lay or very close to doing so. They are already sexed, so you won't get any roosters. They are independent and able to look after themselves in a flock. If you get them from a breeder, they will often be vaccinated against common diseases.

The main drawback with point-of-lay birds is that they are more expensive than younger chickens.

Pros and Cons of Adult Birds

Birds from reputable breeders are proven layers or brooders, so you know what you're getting. They are fully grown and won't need special care. If you want to obtain rare breeds or certain genetic traits, then this is the way to go.

Mature birds may carry diseases or parasites, so this could pose a threat to your existing flock. They might not adapt well to your birds—or to humans. Egg production declines with age, particularly after the birds are older than two years, and dishonest sellers might lie about the bird's age (Rachael, 2020).

Pros and Cons of Rescue Hens

These are hens rescued from culling at commercial farms. If you take on a rescue hen, you are saving her life, improving her quality of life, and might even get a good layer.

On the flip side, rescue hens often have health problems and a short life expectancy. They might not be self-sufficient. Because of the stress of being in the commercial production system, they might not lay as well as other similarly aged chickens.

GETTING YOUR FIRST CHICKENS

Knowing where to get chicks and chickens can be difficult, especially if you are a beginner. Good sources include hatcheries, local farm supply stores, or farmers. Don't use grocery store eggs as these are infertile.

To avoid getting roosters, ask the hatchery before you order whether they sex the chicks or supply only male or female chicks. The mail-order supplier I used the first time did not sex chicks for customers.

Hatcheries

Look online for hatcheries near you—purchasing chicks is just a click away. Finding a local hatchery is imperative so the birds don't need to travel long distances. It's also easier to find rare breeds. While you can place smaller orders in summer, hatcheries only accept orders for larger numbers of birds in winter because too few won't keep each other warm during transit. Although the U.S. Postal Service has been delivering baby chicks for years, things can go wrong, so purchase slightly more than you need to allow for any losses.

This might not work out the way you intend. From personal experience, I recommend you speak to both the hatchery and the shipper so you know what to expect when the chickens arrive. For my first order, I ordered 16 chicks, thinking I might lose two. But the hatchery made the same allowance, which I missed in the brochure, so they sent me 18!

Local Farm Supply Stores

Early spring is the best time to find chicks at local farm supply stores, although breeds might be limited and the sexes unknown. If the store obtained the chicks locally, you might end up with mixed breeds, adding color to your flock later. If the store got its stock from a breeder, they'll know the breed and sex.

It's imperative that you personally observe the chicks and check them thoroughly before buying, as diseases can spread quickly in these stores. Ask about any health guarantees the store may offer.

Adopting Baby Chicks

Over Easter, many people purchase baby chicks. Then they discover the reality of raising chickens and put them up for adoption. These are often listed in your local classifieds or places like Craigslist. You could even find people selling whole flocks if they are relocating or have outgrown keeping hobby chickens.

Contacting a Farmer

The best way to get in touch with a farmer is to ask at your local farmer's market, where someone is likely to know a farmer that breeds chickens or supplies Bantams to people wanting to start raising backyard chickens. If you don't have a local farmer's market or are unsuccessful, consult local directories or search on social media for farmers in your area who breed chickens to sell.

HATCHING EGGS

Choosing Hatching Eggs

Important note: Throughout the process of hatching your eggs, you will not know what sex the bird is until after hatching—and even then, determining sex can be tricky.

Egg quality determines a good hatch rate, so choose your eggs carefully. If you're using eggs from your own hens, this is easy, as you know your hens and what quality care and feed they are receiving. If you are sourcing eggs elsewhere, then you need to check on a few things beforehand.

Check the Source

Buy from a reliable source, preferably a local, independent breeder rather than an online seller. Try and see for yourself how the seller is keeping his flock. Look for active, pasture-raised hens kept with their rooster in a healthy environment and fed an appropriate diet. For postal deliveries, ensure the package is marked "fragile," and ask the post office staff to inform you immediately when it arrives. Store the eggs for at least 12 hours, pointing downwards, before putting them into the incubator (Andrews, n.d.). Twist them several times a day, keeping them cool but not cold and fairly moist.

Check the Age

Eggs should be put into the incubator no more than seven days after they are laid (Andrews, n.d.). Hatch rates will be lower after that as fertility reduces.

Check the Eggs

Eggs that have been sent in the post often have a detached air cell. These eggs usually don't hatch, but try them anyway. Choose eggs that look "normal," i.e., aren't too long, are rounded at both ends and are not pitted. The exceptions are Maran eggs, which are usually rounded at both ends but fine for hatching. Discard cracked eggs. Put the egg above a bright light. The lighter patches on the eggs are porous, and if there are too many, discard the egg as it probably won't hatch (some do, but it's up to you.)

Choose Clean Eggs

Eggs should be clean—dirty eggs can harbor bacteria that may be detrimental to the embryo. Scrape the dirt off with your fingernail, as this leaves the "bloom," a natural protective layer, on the egg. You could also use fine sandpaper, but you might crack the egg or rub the dirt deeper into the shell. Wash the eggs in tepid water.

Step-by-Step Guide to Hatching Eggs

It takes about 21 days to hatch your chicks (Roeder, 2019). The eggs must be kept at the right temperature and humidity and must be turned regularly for the first 18 days, so hatching is a very hands-0n activity (Doug, 2017). In the next section, I'll explain the process so you know what to do if you decide to hatch your own chicks.

Secure Fertile Eggs and Check Starter Feed

Start with at least six eggs to allow for any losses or roosters that you need to discard (Doug, 2017). This should give you a reasonable number of birds to start off with.

Stock up on chick starter feed beforehand. The chicks will need to be fed as soon as they hatch and are placed in the brooder. What you feed them depends on your goals for your chickens, as well as whether they have been vaccinated for coccidiosis.

Setting Up the Egg Incubator

An incubator is a closed unit with a fan and heater to keep chicks warm during incubation—the fan distributes the heat evenly. Consider purchasing an incubator that turns the eggs automatically.

Prepare the incubator about a week before your eggs arrive. Wash it with a 10% bleach solution, followed by warm, soapy water (Roeder, 2019). Rinse it thoroughly. This will ensure that you start off with a sanitized environment for your eggs. Once washed and dried, switch the incubator on to ensure that it maintains temperature and humidity. Place it in a quiet, draught-free place where the temperature is constant.

Maintaining Temperature and Humidity: Hatching Chicken Eggs

The optimal temperature in your incubator is 100.5 °F, although temperatures can range between 99 °F and 102 °F. Make sure that the temperature never drops below 99°F and isn't above 102°F for more than a few hours at most (Roeder, 2019). Use a medical thermometer placed near the incubator to confirm that the incubator thermometer is always working correctly.

Relative humidity for the first 17 days should be 50–55%. This is equivalent to a wet bulb temperature of 85–87 °F (Roeder, 2019). The water channels in the incubator must always be full to ensure the correct humidity. On the 18th day, raise the relative humidity to 70% and maintain this until the eggs hatch.

Use a hygrometer to ensure that the humidity is maintained at the right levels throughout the incubation period. Open the incubator only when necessary, as this can reduce both temperature and humidity. Increase the ventilation as the embryos grow bigger, especially from day 18.

Day 1: Setting Eggs

"Setting eggs" refers to the process of putting eggs into your incubator. Set at least six eggs in the incubator—if you set fewer than this, you risk having only one or no hatchlings, especially if the eggs were shipped (Roeder, 2019). Consider the number of chicks that will hatch together, as chickens are flock animals and need the companionship of others. Place the eggs in the incubator egg tray, with the larger end facing up and the narrower end facing down. Set the temperature to 100.5°F with humidity at 50–55% (Roeder, 2019).

Day 1–18: Turning the Eggs

Incubation begins once the eggs have been set. Rotate or turn the eggs during incubation to prevent the embryo from attaching to the shell—it should always be on top of the yolk. The yolk floats upward on top of the albumen (egg white) towards the shell if the egg isn't turned regularly. This can fatally damage the embryo, which gets caught between the yolk and the shell. Turning the eggs ensures that the embryo is always in the right place.

Turn the eggs at least three times a day, although five is preferable (Roeder, 2019). If you are turning them manually, mark them gently with a pencil—not a pen—to keep track of which ones have been turned. Automatic incubators will turn eggs for you, so you don't need to constantly open the incubator.

Always wash your hands and wear clean gloves before you touch the eggs to avoid transferring skin oils or germs to the embryos.

Days 7–10: Candling the Eggs

Check that the embryos are developing properly around the middle of the incubation period by candling them. This involves shining a light through the egg. White and light-colored shells candle easily, but darker ones need brighter illumination. Use a flashlight or a special light designed for this purpose. Candle a few eggs at a time to ensure that they aren't out of the incubator for longer than 5–10 minutes (Roeder, 2019).

Interpreting Results

- If the egg is clear, it is infertile and there is no embryo. Remove it from the incubator.
- If a red ring is visible inside the egg, the embryo has died. Remove it from the incubator.

- If there are blood vessels inside the egg, there is a live embryo inside. You will see blood vessels about 7–10 days after incubation starts (Roeder, 2019).

Broken or leaking eggs should be removed as they could contaminate the equipment. After candling, return the viable eggs to the incubator and continue with the schedule.

Days 18–21: Pre-Hatching

The embryo takes up most of the egg space by the 18th day of incubation and is getting ready to hatch. Stop turning the eggs at day 18, leaving the larger end facing upwards. The chick will position itself inside for hatching. Keep the temperature at 100.5°F but increase the humidity to 70% (Roeder, 2019).

Day 21: Hatching

Chicks usually hatch on the 21st day of incubation. If the fertilized eggs were cooled beforehand, it might take longer. If the chicks don't hatch at 21 days, wait a few days. Candle any unhatched eggs on day 23 to see whether there are chicks inside. If not, discard them.

Let your chicks hatch on their own without assistance. When all have hatched, lower the incubator temperature to 95°F. The chicks can be moved to the brooder after they have dried off. Make sure that the brooder is ready—the temperature should be set to 90–95°F and there should be water and food inside (Roeder, 2019).

There is a 50% chance that you will get roosters by hatching chicks, as there is no way to determine the bird's sex inside the egg (Roeder, 2019). Have a rehoming plan for roosters if your local ordinances don't allow you to keep them.

CHICK CARE: EVERYTHING YOU NEED TO KNOW

Baby chicks need warmth, feed, and water. Provide comfort, care, and good nutrition immediately, as many chicks never fully recover from a bad beginning. The first few days are crucial, so make sure that everything is properly set up before the chicks arrive. You will need a brooder, heat lamp, and bedding. To ensure that the bedding is dry and the temperature constant, set up the brooder about 48 hours before your chicks arrive ("Caring for Baby Chicks," n.d.).

The brooder is your chicks' first home. It should be warm, draught-free, and comfortable. Ensure there is sufficient space for all your hatchlings. Determining how large your brooder needs to be depends on how long you are going to keep the chicks in the brooder. From my own experience, I would recommend that you start off allowing one square foot per chick. The size requirement expands to two square feet once they are four weeks old and four square feet once they are 12 weeks old. Weather in your area, or other factors may affect how long you keep the chicks in the brooder. Ensure that you have enough space for the final allowance for your chicks depending on how long you plan to have them in the brooder.

Heat lamps can be positioned in the center or at one end of the brooder to keep the chicks warm. Wherever you position the heat lamp, ensure that the chicks have room to move away from the lamp if they get too hot. The lamp must be at least 20 inches above the bedding and 2.5–3 feet from the guard walls. In general, the temperature beneath it should be 95 °F ("Caring for Baby Chicks," n.d.). After the first week, reduce the heat by 5°F per week until it reaches 55°F at the lowest ("Caring for Baby Chicks", n.d.). It is a good idea to consult your hatchery, store, or breeder for their advice about temperature for the specific breed of chick you are raising. A red hue heat lamp is best because it is less stressful to the chicks.

The bedding should consist of absorbent wood shavings. Use larger shavings, about 2–4 inches long (Taylor, 2023). You can use pine but never cedar, as this contains a compound toxic to chicks. Avoid using smaller-sized shavings or sawdust as the chicks peck the ground constantly and might eat it, which can kill them. Larger shavings are too heavy to turn properly so that all the moisture can be removed, and it's difficult for the chickens to move around. The bedding should be 3–4 inches deep so it stays dry and odorless ("Caring for Baby Chicks," n.d.). Remove wet bedding daily, paying special attention to areas around the waterers.

Provide light for 18–22 hours for the first week. Gradually reduce this to 16 hours as the chicks grow until you reach the amount of light they will receive at 20 weeks. Light should be equivalent to a 40-watt bulb for every 100 square feet of floor space ("Caring for Baby Chicks," n.d.).

Allow four linear inches of feeder space per chick ("Caring for Baby Chicks," n.d.). You can use clean egg cartons to dispense feed—these are excellent, and chicks can easily reach them. You can also purchase feeders specifically designed for new born chicks. Then as the chicks grow, you can switch to low feeders or trough feeders.

Heat Lamp Hazards and Alternatives

Heat lamps are usually a particularly hazardous heat source and are expensive to run. The chick's bedding or part of the brooder itself could catch alight. If the lamp is coated with polytetrafluoroethylene (PTFE), which is also used for non-stick pans, it could emit toxic fumes, which could be fatal for your chicks. If you use a heat lamp, ensure that you can monitor it regularly, which is easy to do if you work from home, for example. I personally always use a heat lamp for my chicks because I was advised to do so by my hatchery, and I can monitor it closely all day. Other people never use a heat lamp because of the hazards, so this is another personal decision that you will have to make.

A safer alternative that mimics the temperature underneath a mother hen is a warming unit, or a warming plate, that keeps the chicks comfortable when they're under it. These don't warm up the air like heat lamps do.

I always brood very young chicks inside my house, but you don't have to. Whatever space you use for your brooder, if the location has a cement floor, such as in a garage, basement, or similar place, make sure that the floor is not cold or wet. Put down a few layers of cardboard to insulate the chicks from the floor. Noisy chicks are generally unhappy; if they're quiet, they're comfortable. Provide only as much warmth as they need, and let their behavior guide you.

Introducing Baby Chicks to Water

To provide sufficient water for 25 chicks, fill two one-quart waterers with room-temperature water. Place them in the brooder, away from the heat lamp. Do this 24 hours before the chicks arrive ("Caring for Baby Chicks," n.d.). Some people put stones or rocks in the water until the chicks are two weeks old to eliminate the risk of drowning, but I have never personally had a problem with this.

Wait for a bit before introducing chicks to the feed. Fresh, quality water is essential for chick health. Allow them to drink and hydrate after being in transit. Dip the beaks of each chick into the water to help them find it. Monitor the group for a few hours to make sure that they are all drinking.

Teaching Baby Chicks to Eat

After hydrating your chicks, introduce them to their feed. Use a starter chicken feed with at least 18% protein to give them the energy required for growth ("Caring for Baby Chicks," n.d.). It should also contain amino acids for good development, prebiotics and probiotics to strengthen immunity, and vitamins and minerals.

Place the feed on egg flats, shallow pans, or squares of paper. Add proper feeders to the brooder on the second day ("Caring for Baby Chicks," n.d.). Once the chicks have learned to use the feeders, remove the other options.

Empty, clean, and refill feeders and waterers daily. Raise the height as the chicks grow—they should be level with the birds' backs. The birds' nutritional needs will change as they grow, so adjust the feed after 18 weeks to suit their requirements (Purina Animal Nutrition, n.d.). Transition laying birds onto a calcium-rich feed at 18–20 weeks when they begin laying. For meat birds and mixed flocks, choose a feed with at least 20% protein ("Caring for Baby Chicks," n.d.).

Five Common Chick Problems and How to Solve Them

There are several difficulties you may experience with baby chicks. Here are four of the most common with their solutions.

- Chicks may arrive dehydrated, despite absorbing the yolk prior to hatching. Supply water and teach them to drink on arrival. If they appear to have had a difficult trip, a vitamin/electrolyte solution can help. Add these to the water.
- Coccidiosis is a common intestinal disease caused by parasites that live in the warm, wet conditions in the brooder. It is transmitted via droppings, and baby birds regularly die of it. Symptoms include diarrhea, blood or mucus in droppings, lethargy, restlessness, pale skin, loss of appetite, and general failure to thrive. It can rapidly wipe out several chicks in the same brooder. Prevention is the only way to control coccidiosis. Keep brooders as clean and dry as possible and waterers free of droppings and bedding. Nipple waterers will help to keep the water clean.
- Pasty butt is when loose droppings stick to the down surrounding a chick's vent. Causes include stress, being too hot or cold, or something they have eaten. When droppings dry and crust over a chick's vent, the birds may die if these are not removed. Check the chicks on arrival and daily thereafter. Loosen caked-

on droppings by wiping the vent gently with a paper towel or washcloth dampened with warm water. Don't pull the droppings because you might tear the vent. After cleaning and drying, apply petroleum jelly or triple antibiotic ointment to prevent droppings from sticking to the down. Don't use olive oil. If several chicks develop pasty butt after being under the heat source, it could be too hot—turn down the temperature. You can also mix scrambled eggs with their starter feed. If that doesn't help, change the brand. Adding apple cider vinegar to their water also helps ward off pasty butt. I put apple cider vinegar in my chick's water even if they never present symptoms of any kind as a preventative measure. You only need a few drops per quarter; do not add too much.

- Scissor beak is when the two halves of the beak are misaligned. Most chicks with the condition go on to live normal lives, but it worsens over time. Scissor beak might make it difficult for the chick to eat or drink. If other flock members block her from the feeder, feed her separately. Use a deeper dish for chicks with scissor beak or give them feed with the consistency of oatmeal. Scissor beak requires corrective surgery, but few chicken keepers do this.
- Spraddle leg is a deformity where a chick's feet point sideways instead of forwards. Causes include temperature fluctuations during incubation, hatching difficulties, a leg or foot injury, brooder overcrowding, or a vitamin deficiency. Wet, slippery floors don't help. Avoid using newspaper, plastic, or

other slick surfaces in the brooder. Bind the bird's legs together and do physical therapy until the chick can stand on its own.

Sexing Chicks

Not knowing whether you are raising hens or roosters can be difficult, especially if you cannot keep the latter. There is no sure way to sex young chicks, but there are a few methods you could try.

- Feather sexing is one of the better methods backyard chicken raisers can use, although it isn't accurate for all breeds. It is more reliable if the parent lines are properly sexed. Establish how quickly your particular breed develops adult feathers. In some breeds, the hens have wing feathers, while the cockerels do not. Examine the chicks' wings gently. If the feathers seem longer or vary in size, then the bird is likely a female. If they all appear to be the same length, then the bird is a male.
- Natural markings on your birds can also determine their sex. Rhode Island Reds or New Hampshires have a white spot on the wing of their down feathers, for example, but this disappears as they age. It's important to sex birds while they are still young. Females have a gold gene, while males have a silver gene. Females are often gold or brown, whereas males will be white or other pale colors (Sawyers, 2020).
- Use bird behavior to sex your chicks. As they age, certain behaviors become dominant. Males feed first,

while females hang around in the corners of the coop. Roosters fight and flap at one another.
- Vent-sexing is usually done as a last resort. Get a more experienced person to do it for you if you haven't done it before. Carefully hold the chick upside down so it expels all fecal matter. Apply light pressure to the cloaca so the vent area turns outward. If there is a slight bump, the chicken is male. This is an excellent method of sexing a chick, but it does take practice.

WHEN TO MOVE YOUR CHICKS TO THE COOP

This is a difficult question as there are numerous factors that determine whether your chicks are ready to transition from brooder to coop. If it's mid-summer, the chicks won't need supplemental heat for very long, for example, whereas they'll need it for longer in winter.

Chicks gradually extend the time they spend away from their mother, especially after the first five weeks. Assess your chicks' readiness to leave the brooder six weeks after hatching (Mormino, 2013). There are several ways to do this:

- Look at their feathers. Not all breeds develop feathers at the same age, so it's important to check the feathers in addition to considering the bird's age. At six weeks, the birds should be fully feathered so they can regulate their own body temperatures.
- Consider the ambient temperatures and time of year. The behavior of your birds is a much better

indication than brooder temperature of whether they are ready to move to the coop. Remove the heat lamp or heat source once they are spending very little time near it. They won't need it if they are ready to move. If outside temperatures are averaging 65°F and the chicks are at least six weeks old, they can be moved (Mormino, 2013). If they are huddled together and noisy after moving, they are likely cold, but if they are behaving normally, then they are comfortable.

- Check the set-up of your coop. If your chicks need extra heat once they've been moved, is there electricity available? Use a heat plate rather than a heat lamp if temperatures drop once your chicks are in the coop and they are cold to avoid the risk of fire. Alternatively, wait until outside temperatures are warmer before moving them.
- It's also essential to ensure that the coop is as predator-proof as you can make it (see Chapter 3 for guidance on how to predator-proof your coop.) If not, predators might find your chicks really quickly.
- How many chicks are there? The more chicks, the more chance they'll be able to keep one another warm with their body heat.
- Do you have any older flock members? If so, then it's better to wait until the chicks are closer to their mature size before moving them so they aren't hurt by bigger birds. Integrate the two groups slowly to avoid conflict and stress.
- Take into account the stress of the transition. This is a very big move in the life of small birds, but there are things you can do to relieve it:

- Keep them in the coop for a few weeks to get used to their new home before letting them go into the run. Chicks that aren't given this time to adjust might not return to the coop after dark, which is stressful for you and hazardous for the birds.
- Block access to the nest boxes so the chicks can't hide there. They will tend to do this but don't allow it as sleeping chickens poop. This will become an issue when they start laying eggs, as the eggs will become contaminated. Once the chicks are 17 or 18 weeks old, the boxes can be opened again (Mormino, 2013). If the coop is being used by egg-laying chickens, you can close off the boxes temporarily with plywood or cardboard in the afternoons when the egg-laying is finished for the day to keep the boxes clean. However, none of the Crazy Chicken Ladies that I know do this, nor do I, because of the time commitment. If you do chose to block boxes for part of the day, open them up first thing, allowing laying birds to use them while the younger birds are excluded.

PRODUCTION CYCLE OF LAYER CHICKENS

Hens usually start laying eggs from 18–20 weeks of age. Initially, egg production and size will vary, but they will settle into a routine, and egg sizes will stabilize, although they will lay fewer eggs. Productive birds usually lay an egg a day—it takes around 24–26 hours to create an egg (Biggs, 2018). Hens stop laying when they molt, usually in the fall. They also lay fewer eggs if there are less than 12 hours of light a day (Biggs,

2018). Provide supplementary light to keep your hens laying for longer.

Hens are most productive at around 30 weeks. Egg production gradually reduces after they reach the age of two years, with hens laying around 80% of the eggs they did in their first year. In the third year, this drops to 70%, and in the fourth year, to 60% (Biggs, 2018). Older hens produce fewer eggs, but they may be larger.

Chickens live for an average of 3–7 years, although they might reach the grand old age of 12 if they are carefully looked after and kept in ideal conditions, safe from predators (The Editors, 2021).

WHEN HENS STOP LAYING...

If you are keeping chickens as farm animals rather than pets, they might contribute to the homestead when they stop laying. Keep an eye on them so younger hens don't harass them.

Older chickens can control pests, including mosquitoes and flies, as well as garden weeds. They can watch for predators—and are usually more effective than younger hens. They continue to provide garden manure. They are better broodies than younger hens, as they happily sit on the nests for long periods. They are good, experienced mothers.

You can certainly allow your chickens to die from old age. However, if you want to limit the size of your flock and have a consistent number of actively laying hens, eventually they will need to be slaughtered. Or, if you have no alternative for unwanted roosters, they will need to be slaughtered. And of course if you are raising chickens for their meat you will need a process for slaughter. Chickens can be humanely dispatched by wringing their necks or quickly cutting their throats with an ax. Many people use what is known as a poultry cone, or restraining cone. The cone holds the bird upside down, with its head poking out of the bottom of the device, to allow for the animal to be more easily bled. You can use a pair of large shears to remove the head quickly, in one motion, thus dispatching the animal quickly and humanely. Whichever method is used, you can hypnotize or calm the chicken first. Hold it breast-down by the legs (if you aren't using a cone) and wave a piece of chalk in front of its eyes. Slowly draw the chalk out from the beak by 12–18 inches (The Editors, 2021). The chicken will focus on the chalk and won't flap or move. Alternatively, lay the bird on its side and tap near the point of the beak without touching it and then four inches in front of the beak. Repeat until the bird calms and stays still (The Editors, 2021).

SUMMARY

Before acquiring your chickens, assess your chicken-keeping goals. Find breeds that suit your requirements, whether you want egg layers, meat, or family pets and interactive experiences for your children. This will determine what age the chickens should be when you purchase them, although there are no hard and fast rules.

Bear in mind the chicken life cycle, where the embryo is formed in a fertilized egg. This hatches as a chick, which grows to become first a pullet or cockerel and then an adult hen or rooster.

Although it is usually cheaper to buy fertilized eggs or day-old chicks, these options do require special equipment like incubators, brooders, and heat lamps, so factor these into your calculations. Chicks and larger birds can be obtained from different sources, including online by mail order. Hatching out your chicks is a 21-day affair, after which they will need to be moved into a brooder and then into the coop about six weeks later.

When hens are too old to lay and cannot fulfill their roles around the homestead, they might need to be humanely dispatched.

In the next chapter, you'll find out everything you need to know regarding chicken coops, from building and fencing them to predator-proofing them—and even how to manage roosters, assuming you want male birds

CHAPTER 3
Flocks in the Coop

Regard it as just as desirable to build a chicken house as to build a cathedral.

— FRANK LLOYD WRIGHT

Building your own coop can be an interesting experience. Some coop owners opt to have only one entrance—usually a smaller one that the chickens can easily use. The same is not true for the chicken keeper, however. After months spent crawling in and out on your hands and knees when you enter the coop to collect eggs and do your chicken chores, you will most likely decide to alter the coop so you can enter and exit through a human-sized door.

CHICKEN COOPS 101

It's not difficult to build a chicken coop. Your chickens won't need lights and running water, but they do need shelter from the elements. The structure must be waterproof - wet chickens aren't happy birds. You also need to ensure that your coop is predator-proof—a safe place that protects your flock. So, what should you consider before starting construction?

- The coop's location will enhance its benefits. Build it on high ground if possible because this will reduce flooding, control mud, and prevent water from building up. Build the coop close to the house or in a place where there's a lot of human activity to deter predators. Build it in an open area free of vegetation so there's nowhere for them to lurk undetected. The coop should get enough light to encourage egg laying, so ensure that it faces south. There should also be some shade to provide relief during very hot weather.
- The coop must be large enough to hold your flock comfortably. If your chickens are overcrowded, they will become unhappy, aggressive, and antisocial. Winter can be problematic if the season is particularly cold, as the chickens could get bored. Allow sufficient space in the utility areas for several birds to congregate, as chickens want to do things together.
- There are several flooring options but not all are created equal. Plain, unfinished plywood covered with a layer of wood shavings is recommended. You

can use linoleum as a "poop deck" on the surface to protect the wood, but this will prevent moisture from draining through the bottom. Consider how wet or dry your local climate is when considering this option.

- The coop should be cool in summer and warm in winter. Good ventilation is essential. Install air vents near the ceiling above the roosts. Cover them with hardware cloth to prevent predators from getting in. Don't use chicken wire as predators can easily break it.
- Chickens want secluded places to lay eggs. Nesting boxes need three sides and a roof and must be elevated off the ground. Boxes for medium-sized chickens should be 12 inches wide and 18 inches high, while larger hens will need bigger ones (Cooper, 2022). Provide several, so hens don't lay eggs on the floor, where they might become contaminated. Line each box with bedding such as straw and change it regularly to ensure it remains clean and parasite-free.
- Chickens shouldn't roost on the floor. Your coop should have roosting bars for the birds to use at night when they are sleeping. There needs to be enough room so the roosting area doesn't become overcrowded, and hens can have personal space if necessary. Allow about 8–12 inches per bird (Guest Author, 2020). Roost bars should be 2–5 inches in diameter and placed at least 18 inches from the ground (Telkamp, n.d.). Don't use metal or plastic as they may become slippery, causing foot

problems. Use sturdy branches, rubberized poles, or lumber.

- Your birds will want some outside space, regardless of whether they are free-range or not. A chicken run creates a place where they can enjoy fresh air, sunshine, and worms. They must spend some time in the sun to absorb vitamin D. Provide them with a sheltered spot in the run where they can spread their wings to catch the rays. They love doing this communally, so the space needs to be large enough to accommodate all your birds.
- Your birds will need to dust bath to keep their feathers in good condition. When preening, they spread oil from a special preen gland near their tails onto their feathers. Dust bathing removes this oil and any parasites from their feathers and skin. Choose a dry, sheltered spot for dust bathing and ensure that it's always clean. Dirt, sand, or peat are the best options for dust baths.
- Secure your coop, and your flock, from predators. Raise it off the ground by at least 8–12 inches, allowing chickens to walk beneath it (Lesley, 2020). With raised coops, the wood is less likely to rot. A dirt floor with wire underneath to keep out digging predators is a good alternative and will prevent rodents and snakes from getting into the coop.
- All coop doors and windows must have secure locks to prevent predators from breaking in. A secure door large enough for you to enter and exit when collecting eggs or cleaning the coop is essential. Use spring-loaded locks or padlocks so predators can't

unlock them. Automatic chicken doors that work on timers can be purchased for coops. Remember to check that you won't be locked in if the door closes unexpectedly—it's surprising how often this happens.
- Consider whether the coop will need electricity. Providing low-wattage lighting in winter will keep egg production up. Having electricity will also enable you to provide supplemental heat if needed.

BUILDING THE COOP

Preparing the Ground

Don't start building after heavy rains as the ground will be soft and unstable. Clear the area of rocks and sticks and cut back heavy foliage around it. Remove or relocate sheds, rock piles, or dark and shaded areas that are potential predator hide-outs.

Choose Your Plan

There are numerous chicken coop plans available online. Choose one that suits your flock size and chicken breeds. The coop need not be elaborate, but if you want to get creative and have the funds to do so, go nuts!

Build Your Coop Frame

The frame depends on the coop size. Build it carefully, making sure that it is sturdy and secure. You can use second-hand and repurposed materials for the rest of the coop but splurge on quality lumber for the frame to avoid paying for it later with an unstable, leaky, drafty structure that might collapse in time. Double and triple-check all elements, including measurements, angles, and fastenings, making sure they are all correct. You can paint the exterior parts of the frame to protect them from the weather before assembly.

Install Coop Walls

Add wall panels to the coop, together with at least two panels for the roof (Lesley, 2022). Have the plywood cut at the lumber or timber store. Fasten the panels to the frame securely, ensuring it is flush along the edges with no gaps that could allow drafts or predators in. Cut the vents and cover them with hardware cloth so they are predator-proof. Make sure vents above the perches can be closed during winter.

Lay the Floor

While a dirt floor may be recommended in some coop plans, consider a wooden, elevated floor, as this will make the coop safer and drier. The floor need not be entirely even, but it must be level with well-secured, stable boards. The wood shouldn't have knots or holes, as predators can get in through these. Put hardware mesh underneath the floorboards to deter snakes and burrowing predators. After the side panels are installed, screw in the floor, securing it to the coop frame.

Add the Doors

Your coop needs two entrances—one for the birds and one for you. You can build slightly larger bird entrances and squeeze through them or make an entire wall panel removable so you can get in. Both entrances must be accessible and need to be secured to keep birds in and predators out. Your entrance could be closed with hardware mesh to make the coop cooler in summer. This can be similar to a screen door that can be closed and securely locked while the wooden door remains open. Use heavy gauge hinges and nails rather than staples to guarantee that the coop is sturdy and predator-proof.

Add Nesting Boxes and Perches

These can be basic, as chickens aren't fussy. Allow one nesting box per three hens, and add more if your breed is broody (Lesley, 2022). You can also consider brooding boxes. These have a latch that can be opened from the outside to collect the eggs—store-bought, premade coops often feature these.

The chickens will make up their own minds, though. Hens often use the same nesting box to lay eggs, so you could find multiple eggs in the same box most of the time. One chicken keeper on social media tells of how all her hens want to use the same box to lay their eggs. One morning, when she went into the coop, one of the hens was screeching at the one laying as though to tell her to hurry up because she was taking too long. You can put golf balls or ceramic eggs into the nest boxes to encourage the hens to use them all.

For perches, a 2x4 will suffice as long as it's higher than the nest boxes (Lesley, 2022).

Build the Run Frame

The same principles apply for run frames as for coop frames. The run frame will only need to support chicken wire or hardware wire, so it doesn't need to be as robust. Don't skimp on construction, though, as it still needs to be secure and predator-proof.

Add Run Fencing

Hardware wire is preferable to chicken wire, as it's better for keeping predators out and chickens in. While you can use chicken wire above three feet from the ground, use hardware wire below and sink it at least six inches into the ground to keep out burrowing predators (Lesley, 2022). It might be more expensive to construct the run this way, but it will be worth it in the long run.

Add Amenities to Your Coop

Use waterers to keep the water clean and allow one per 3–4 chickens (Lesley, 2022). If you use a waterer that is suspended off the ground and has several nipples to give multiple birds access, it will be easier to keep the water clean.

Get a feed trough long enough that all chickens can eat at once. Have enough wood shavings to cover the floor at least 6 inches deep, and put a few handfuls into the nest boxes as well (Lesley, 2022). Change the bedding regularly.

Finally, check that everything is tight and secure. Conduct regular maintenance checks to look for any warping or damage. A well-built coop ensures that your chickens are healthy, happy, and secure.

FENCING

You can't always keep a good chicken in. Sometimes a rooster will leave your coop, apparently for new pastures, only to show up at your neighbor's place. Other hens will fly over the fence, come back to your porch to lay their eggs and go back over the fence again. But you need to keep predators out as they can decimate your flock in no time, especially if you have few birds.

Ask these questions to establish how high your fence should be.

1. Does your area have a high population of predators? Common chicken predators include foxes, coyotes, possums, wolves, weasels (which love eggs), rats (which also love eggs), fisher cats, snakes, wild or domestic cats and dogs, birds of prey, mountain lions, bobcats, and skunks. You might need a fence as high as 5–6 feet to keep these predators at bay (Mike, 2021).
2. How does your flock behave, and do the birds like flying? If your flock are active birds that often stray beyond their chicken coops, then you might also need to look at a six-foot-high fence (Mike, 2021).
3. Do you live in a busy area with lots of traffic? If this is the case—and especially if you live on a busy street—then your fence should also be six feet high (Mike, 2021).

A lower fence (2–4 feet high) offers moderate protection against predators and will deter your birds from escaping (Mike, 2021). But it's not ideal, as even docile, lazy hens will fly over it if they really want to. It's better than nothing, though.

Chicken fences should be between four and six feet high—a height that will keep most (but not all) predators out and chickens in. But there are still predators that will find a way to access your chickens—and chickens determined to explore new pastures. A sturdy fence over six feet tall will provide excellent security for your chickens (Mike, 2021).

There are several options for fencing materials:

- Plastic poultry netting is affordable and quick to install, but it won't protect your flock adequately from determined large predators like foxes, coyotes, or raccoons. However, it is a good option for fencing off parts of your yard you don't want your chickens to access while they are free-ranging.
- Metal poultry netting is economical and will keep most predators out.
- Hardware cloth is an excellent material to use, but ensure that the weave is tight enough to keep small predators like rats and snakes out of your chicken yard.
- Use welded wire to seal up loose ends in your chicken yard, shore up your chicken fence, or install a custom wire chicken fence.
- Chain link fencing is an excellent choice for securing your chickens and will make predators think twice. Consider using a privacy screen so they can't see your birds.
- Tall stockade fencing is the best option as it will keep out most land-based predators, although it can be expensive.

Why Do Chickens Leave Your Yard?

You could be forgiven for thinking that this is because they crave new adventures, like roosters that hitch rides into town on the backs of their owner's trucks. But the real reason is that they are under some sort of stress at your place.

Your flock needs plenty of space. They want to stretch, scratch, explore, and peck without boundaries. The coop should be roomy, clean, and pest-free. Chickens need plenty of grass for forage. If your chickens were recently surprised by a predator, that's reason enough for them to consider a relocation. The best way to prevent them from pulling a Houdini is to ensure that they are in a calm, secure environment with plenty of food and water and a clean, safe coop.

PREDATOR-PROOFING YOUR CHICKENS

If you don't know what predators are in your area—but hopefully you do—then it's important to find out. One predator to watch is birds of prey, especially hawks and eagles that swoop down to catch your chicks unexpectedly. Predators are one of the biggest challenges chicken keepers face, and the bigger they are, the more extreme your defenses need to be. A friend of mine in Valdez, Alaska, who keeps chickens has resorted to having electric fences around her coop—separated from the chickens, of course—because of the large bear population there.

Here are some basic tips for dealing with predators.

- Never allow your birds to roost outside. They are most vulnerable when they are asleep, and many predators are active at night. Train your chickens to return to the coop as soon as they start using it.
- Chicken wire won't keep chickens safe. It's intended to control chickens, not keep predators out. Hawks and eagles can reach in and take a bird with their talons.
- Hardware cloth, as mentioned previously, is an excellent predator deterrent. Dig a 12-inch trench around your coop and bury the hardware cloth in it. Do the same for dirt floors (Mormino, 2013). Create an apron of hardware cloth out from the perimeter of the run to deter burrowing predators.
- You must ensure that you close *every single* hole, no matter how small, with hardware mesh. When I built my first coop, there were a few small holes the size of quarters and dimes in the bottom of the coop. When I moved my first brood of guinea hens out to the coop, half of them (nine chicks) disappeared on the second night they were in the coop. The doors and windows were completely secure. The only access was those small holes. Nine chicks in one night. Be sure to put mesh over *every* square inch of the coop!
- The coop must be roofed or covered with hardware cloth to stop flying predators in their tracks. Netting will deter birds of prey but not climbing predators.
- Close the access doors to your coops and chicken runs at dusk. Remember to check that the run is empty before securing doors for the night. Do this

even if you have automatic doors—predators can get stuck inside, waiting for your birds in the morning.
- If your coop has windows, screen them with hardware cloth so predators can't get inside if they are open.
- Don't leave food out overnight, as this attracts rats and mice, and store your feed well away from the coop. Lock it up and secure it.
- Manage your chickens so they aren't inadvertently attracting predators. Collect eggs daily so there aren't any left in the coop. Clean up scraps your birds don't eat.
- Predators can be scared off with motion-sensored lights, hot wires around the coop if you have large predators like bears, or electrified chicken netting.
- Use other animals to guard your flock. These guardian animals include certain dog breeds, roosters, other poultry like turkeys and guinea fowl, or livestock such as donkeys, goats, sheep, or cows. Attracting crows and bluejays to your chicken-raising area will also help to deter predators.

RAISING FREE-RANGE CHICKENS

I always encourage free-ranging your chickens during the day, although it might not be an option if you are in an urban area. If you are, follow the recommended sizes of runs per animal. Even if you have some acreage to work with, you may still decide not to free-range your chickens. If you do, be aware that they get stuck in strange places at times. If you can't find one and there's no evidence of predation—no feathers or blood anywhere—then look for places where she might be stuck.

Pros and Cons of Free-Ranging

There are several positive spin-offs to free-ranging your chickens.

- You get tastier, more nutritious eggs with much darker yolks than from hens that don't forage for their food. For best egg quality, chickens should be eating pellets, forage, and insects.
- The birds won't consume as much chicken feed when they are supplementing their diet with forage, saving you money.
- They'll find their own dust bath areas and grit, reducing the demands on you.
- Chickens love foraging. It keeps them busy and active and improves their health because they're less likely to become obese, which is a problem for urban chickens.
- They'll naturally control pests, provide fertilizer, and till your ground.
- You won't need to worry about crowding and all the birds getting enough food and water.

Having said that, there are some disadvantages to having chickens free-ranging in your yard.

- You will need to protect them from predators while they are outside, as it's very heartbreaking to lose chickens that have bonded with you and your family.

Having a run is not enough—you need to secure the coop and have proper fencing.
- Chickens are curious, but your neighbors might not appreciate having feathered trespassers. If they complain, you might need to keep your chickens in their run. Chickens may be at risk if they encounter dogs that aren't chicken-friendly.
- Eggs might mysteriously disappear as the hens lay them in odd places. If obtaining eggs is one of your chicken-keeping goals, then this might be an issue.
- They will till your garden whether you like it or not —including areas where you'd prefer them not to. They'll eat everything, including your favorite flowers.
- Unless you've trained them, they might not be willing to roost inside the coop at night.

Teaching Your Chickens Where to Lay Eggs

Confine the flock to the yard when they're ready to start laying. You'll know this because their combs and wattles go dark crimson, their legs turn paler, and they squat when you walk up to them. This is to allow the rooster, if you have one, to fertilize the developing eggs. Put ceramic eggs or golf balls into the nests so the hens see them. This shows them where to lay. Once the routine is properly established and they know the ropes, free-range them again.

Training Chickens to Respond to Your Calls

Put their feed into a specific colored bucket. They will associate that bucket with food and will come to you when you have it out. If they are free-ranging and it's coming up to roosting time, go out into the yard with the bucket, and they'll arrive because they want to see what's in it. This makes it easy to get them into the coop before dark.

Free-Range Compromises

Chicken tractors are an ideal solution for those who either don't want to free-range or where local laws don't allow it. This covered run on wheels can be moved from one part of the yard to another, with the chickens fertilizing as they go. The birds will eat whatever insects they find. This keeps them out of areas you'd prefer them to avoid and offers some predator protection while they're outside.

Create a covered, fenced area large enough for the whole flock. This will provide many free-ranging benefits, but the birds will be safe—and so will your garden. Replant the grass from time to time so they always have forage available. The chickens will eat all the vegetation and insect life in the enclosure, so this needs careful planning.

ROOSTERS: THE BIG QUESTION

This is the ultimate chicken lady deliberation: To have a rooster or not. There are definite advantages to having a rooster—but there are some drawbacks too.

Benefits of Roosters

- Good roosters are amazing bodyguards for your girls. Roosters are vigilant, constantly watching for predators and alerting the flock to run before the

threat arrives. Roosters will remain on guard after the hens have scattered.
- Because the rooster attends to security and leads the flock, the hens are more relaxed. As the rooster is automatically the flock leader, there is less friction between the birds, so you get more eggs.
- Roosters look for delicious morsels for the hens and call them to share the bounty, ensuring that they get the best nutrition.
- Did you know that roosters can dance? The waltz is actually an act of dominance, with the females running off in acknowledgment that he's the boss. It is entertaining to watch.
- If you have a rooster, you can hatch out your own fertilized eggs and get more chicks. You don't need a rooster if you just want eggs for consumption, but if you want to raise more chicks, then you do.

Disadvantages of Roosters

- Roosters spend most of their time mating with all the hens in your flock—and do this numerous times a day. This becomes very tiresome for the females. Hens lose neck and back feathers. Try using a chicken saddle, a vest-like garment, for hens to wear over their backs so they don't lose so many feathers—or all of them.
- Roosters crow, usually at first light. They crow at other times too, sometimes apparently to annoy the neighbors, but really to let everyone know who's boss.

- While roosters are wonderful predator deterrents, this isn't so great when you go on vacation and someone comes to look after your hens. Roosters bite hard. They have spurs on their legs that can tear flesh. When choosing a rooster, you may want to find one from a less aggressive breed.
- Some roosters are chronically jealous and might even kill chicks if they feel a hen is paying too much attention to her offspring. Thankfully, this doesn't happen often.
- It is illegal to keep roosters in some areas, so check with your local authority before acquiring one.

PREPARING FOR STORMS AND SEVERE WEATHER EVENTS

Most chicken keepers don't routinely prepare for severe weather, but it's a good idea to have an emergency plan in case the worst happens. Here are some tips to help you get started.

- Determine the severity of the storm and prepare accordingly. Make early decisions concerning protecting your flock—it's better to be over-prepared. Give yourself plenty of time to prepare if severe weather is forecasted.
- Check feed levels and medical supplies and move the former to a dry, high-lying place. Allow 24 hours to do yard clean-ups and coop prep (Brahlek, 2022).
- Make a list of emergency contacts, including:
- your own vet and an emergency vet

- details of boarding kennels where you might need to home your chickens if the coop is destroyed or your property flooded
- contact details for friends, neighbors, and colleagues who could home them temporarily.
- Assemble a chicken first aid kit. This should include:
- Epsom salts for bumblefoot soaks
- regular first aid items (gauze, bandages, cotton swabs, bandage tape)
- Corid for coccidia
- Blue-Kote for small wounds and feather pecking
- styptic powder to stop bleeding quickly
- sterile gloves
- vet wrap, a self-sticking bandage for wound care
- super glue for broken nails and emergency "stitches"
- nail clippers for broken nails
- tweezers
- dried grubs to attract an injured chicken and keep it distracted during care
- Keep a toolbox in an accessible location that contains tools and supplies for repairing the coop or your home. These might include:
- tarps
- drill and screws
- flashlight and extra batteries
- zip ties
- rope and bungee cords
- hammer and nails
- duct tape
- work gloves

- The coop is usually the perfect place for your flock to wait out a storm. Cover some ventilation vents but leave enough open to give your birds air. Secure any branches or loose items around the coop or nearby. Move feeders and waterers into the coop.
- However, if you feel the coop would not be the best place for your birds during the storm for any reason, consider where your flock can be housed as an alternative. Prep the floor if you are moving the birds to your basement, garage, or house. Lay down a tarp or drop cloth and add a layer of bedding. Cover or move anything that should stay clean. Ensure that your girls have elevated roosts where they can keep their feet dry. Windows should be boarded up (a windowless space like a garage without an automatic door is better.)
- Before the storm hits, secure your flock in the place where they'll be during the event. If your flock is staying in the coop, check that the locks are secure and will remain so.
- You might not be able to take your entire flock with you if you need to evacuate but take them if you can. Use a dog crate or totes with ventilation holes. Don't forget their food, water, and first aid kit.
- Stay put during the storm. Going out to check on them might be dangerous for you and them and could add to their stress.
- Once it is safe to leave your home after the storm, go out and check on your flock, exercising care.
- First, assess the exterior of the coop for damage or places where chickens might have escaped. If you

have escapees, round them up and remember that some might take a few days to find their way home.
- Fill up waterers and replenish feed. Check each chicken for injuries and treat them as required. If there are severely injured chickens, quarantine them and limit their movement. If water has entered the coop, move your flock to a dry location as soon as possible to keep them away from potential pathogens. Dry them with hand towels. You can also use a heater or hair dryer.
- Remove any storm debris that could injure your hens.
- Repair the coop, starting with the roof, and use hardware cloth to repair gaps and holes. Make sure the coop is secure before rehoming your chickens there. Take as much time as you need.

Storms can have their funny side. One chicken keeper thought her hens had stopped laying—until she and her husband removed the tarp they had thrown over the coop before a storm hit. Turns out the hens were laying on the tarp. They discovered this when they started taking the tarp down, and the eggs cracked on their heads as they fell.

SUMMARY

It's essential to have a chicken coop to keep your birds safe from the elements and any local predators, especially at night. Ensure that the structure is robust and properly constructed, paying special attention to securing the doors, windows, and vents so predators can't get inside. Add a run to create an outdoor area for your flock to enjoy some outside time. Proper fencing is essential, not only to keep chickens in but also to exclude predators.

Keeping free-range chickens has lots of benefits—and a few drawbacks, which can be overcome by employing some useful techniques. Roosters can be beneficial if you are allowed to keep them in your area, but they can be detrimental too. It's up to you as the chicken keeper to decide.

Always be prepared for a severe weather event and ensure that your flock is safe for the duration.

So now that you know how to house and protect your chickens, the next chapter will focus on how and what to feed them.

CHAPTER 4
How—and What—to Feed Your Flock

Pride's chickens have bonny feathers, but they are an expensive brood to rear. They eat up everything, and are always lean when brought to market.

— ALEXANDER SMITH

Chickens are rather like goats—they will eat anything they can get their beaks into. They are omnivores, eating both animals and plant material. In the yard, chickens will find many things to eat, from earthworms, insects, and slugs to the tips, flowers, and seeds of plants like clover, buckwheat, Kentucky bluegrass, and weeds. You might even see a rooster catching a mouse for its chicks. Chickens may also try eating larger animals, including skinks, small toads, and even snakes, before establishing that these are not ideal foods.

HOW TO FEED CHICKENS

Chickens have different dietary requirements depending on whether they are being raised to produce eggs or for meat. Backyard egg-laying chickens can be fed household food scraps in addition to their feed. Feed must be monitored to ensure that your birds don't overeat, as this can be fatal.

Meat birds raised indoors or free-range need high-protein feed so they reach their goal weight efficiently. You might need to decide whether to vaccinate your meat birds, as this will mean that you need not feed them medicated feed. Birds with significant outdoor or yard access generally eat a more balanced diet comprising natural foods as well as feed.

The 90/10 Feed Rule

Certain foods, including table scraps, mealworms, and scratch grains, are chicken treats. Provide these in moderation, following the 90/10 rule, where 90% of the diet should consist of complete layer feed, with the remaining 10% being composed of "treats" (Biggs, n.d.b).

What to Feed Chickens

Chickens need 38 specific nutrients in the right quantities (Biggs, n.d.b). Give backyard chickens layer feed in both summer and winter. This is specially formulated and balanced to ensure that they receive the right nutrient mix for optimal health and continued egg production. Protein and calcium are important, but the birds also need vitamins and minerals, which are included in commercial feeds.

The best is to choose a complete starter-grower feed from day 1 until 18 weeks and one complete layer feed for laying hens thereafter (Biggs, n.d.b). Introduce treats after hens have laid their first eggs—the birds will need all the nutrients in their starter-grower feed to ensure good growth and optimal health. Directions from different feed manufacturers vary. Check the feed bag and follow the instructions. Your birds will eat a 1/4 pound of complete feed daily, which equates to around 1/2 cup. With the 90/10 rule in place, treats shouldn't be more than 2 tablespoons a day (Biggs, n.d.b).

Your chickens will need grit—sand, gravel, and coarse grit—to grind up foraged foods. Grit is kept in their gizzards.

Oyster Shell Supplements

A hen's egg contains 94–97% calcium carbonate, all derived from her body. In just one year, a laying hen provides 20 times more calcium for her eggs than that in her own bones (Roberts, 2019). Oyster shell supplements are ground-up oyster shells, and they're extremely rich in calcium, which is essential for laying birds. Giving your layers this supplement lowers the chances of broken bones and eggs. It also strengthens and improves chickens' blood vessels, immune systems, and cardiovascular functioning.

Signs of calcium deficiency include soft (or no) eggshells, broken bones, joint injuries, a drop in egg production or failure to start laying, stiff legs or lameness, as well as behavioral problems like excessive pecking or hyperactivity.

Supply the oyster shells separately to the feed, especially if you have hens that are not yet laying or a mixed flock of both layers and meat chickens. The chickens will only eat what they need.

Pasturing Meat Hens

If you are raising meat hens on a free-range pasture, they won't grow as fast as those in coops eating broiler rations. The meat is dense yet tender, and the omega-3 content is higher than in chickens that are confined and fed grain-based feed.

Kitchen Scraps and Other Treats—What Can You Feed Your Chickens

Chickens will literally eat anything—including chicken. (It's not recommended that you give them that, though, as it might encourage cannibalism.) Feed your chickens the following kitchen scraps as treats only:

- beef and pork scraps, including the tendons, gristle, and fat
- congealed fats and oils
- cooked rice, pasta, and vegetables
- dairy products, including milk, cheese, and yogurt
- fish and fish skin with no bones
- fresh fruits (there are some prohibited ones—see below)
- oyster shells and eggshells for calcium
- stale bread and crackers (without mold)
- wilted salad greens

It's not necessary to cut up the scraps before feeding your chickens, although you can if you want to.

Other foods chickens love include:

- herbs such as lavender, mint, oregano, parsley, cilantro, thyme, and basil
- flowers like daylilies, hostas, daisies, roses, coneflowers, and ferns
- vegetables like lettuce, beets, broccoli, carrots, kale, chard, squash, pumpkins, and cucumbers

Remember to fence them off from your garden so they don't eat all your fresh produce before you've had a chance to harvest it for your family.

Providing Water

Always ensure that your chickens have plenty of fresh, clean water available as they need this to digest their food.

Emergency Feed

If you run out of feed, you can give your chickens eggs. Hard boil and chop them up or scramble them first. Don't do this on a regular basis, though, as it could encourage them to eat the eggs they lay. Your chickens can go a day or two without feed—or even longer without any problems, as long as they are getting some kitchen scraps.

Making Your Own Feed

Buy, mix, and even design your own custom feed for your flock if you wish. If you are on a large property, you might choose to grow the grains, seeds, and other ingredients found in commercial chicken feeds.

Vitamin and Mineral Supplements

Free-range flocks are likely getting enough vitamins and minerals, but supplementation is recommended for birds that are confined and for laying hens. Egg-laying hens need vitamins A, B2, D, E, and K (James, 2022). These minerals boost their health and promote good egg production. Chickens need minerals like calcium to ensure strong bones, good hatching, and hard eggshells. Calcium is found in oyster shells, limestone, and bone meal. Another important mineral is zinc for feather development and to prevent bone diseases. Manganese stops perosis, a condition that affects poultry. Cobalt ensures that baby chicks grow properly (James, 2022).

Your flock should get vitamins every day. Chicks need vitamins and minerals to ensure that they grow well and develop properly. Niacin, a water-soluble vitamin, is essential for ensuring that baby chicks have straight, strong legs and healthy mouths and tongues. Choline promotes timeous hatching and good growth thereafter. Folic acid ensures that chick feathers develop properly and prevents anemia (James, 2022).

Free choice supplements—ones that aren't in feed but set out in a separate container for your chickens to eat voluntarily—include grit, oyster shell, certain herbs (as mentioned previously), and protein booster supplements to improve feather formation and development. Chickens need more protein in winter.

Mealworms contain 12 of the 16 elements of living tissue that aid tissue growth, especially feathers—and chickens love them. They need very little—one tablespoon 2–3 times a week will get them through the molt (Sextro, 2019). Black soldier fly larvae are another excellent protein supplement, promoting feather growth, improved energy levels, and harder eggshells. Grubs cost more but provide your hens with 50% more calcium than mealworms (Sextro, 2019).

Bone meal contains calcium and phosphorus, while pumpkin seeds contain magnesium, one of the essential minerals for chickens. Mix the seeds with the feed to ensure maximum absorption. Peanuts and cashew nuts, together with dried fruits, are also excellent magnesium sources.

Natural Supplements

Apple cider vinegar is invaluable during the winter or the rainy season when respiratory ailments threaten. Add one tablespoon per gallon of water every second day as a preventative (Sextro, 2019). Use an organic product for the best results. The vinegar aids digestion and helps reduce internal parasites.

Garlic is a natural antibiotic and immune booster, relieving respiratory infections and killing parasites. Add fresh cloves or garlic powder to feed. Alternatively, crush a garlic clove and add it to a gallon of water. Give garlic to your flock once a week (Sextro, 2019). Offer small chicks crushed garlic so they develop a taste for it.

While sprouted grains are on the list of chicken treats, they provide them with vitamins B and C, folate, and fiber and help to boost their immune systems (Sextro, 2019).

Added to feed, molasses is high in iron and minerals. It's usually used with a probiotic powder to ensure that it sticks to the feed. Use ¼ cup of molasses per gallon of feed and mix well (Sextro, 2019).

Oregano oil is a valuable supplement that commercial chicken farms use. It is a natural antibiotic and is used as an alternative to conventional medication. Oregano essential oil is very concentrated, so don't give it to your chickens in their feed. Adding fresh or dried oregano to their diets is very beneficial as it strengthens their immune systems and helps prevent common diseases like salmonella, infectious bronchitis, avian flu, and E. coli. Chickens love fresh oregano. If you grow this herb, dry it and add it to their drinking water.

Specially Packaged Supplements

These include probiotics that boost the immune system and aid digestion. They come in powder form and can be added to drinking water.

Chickens lose electrolytes when experiencing stressors like heat, overcrowding, predators, and others. Electrolytes rebalance their physiology. Add the powder to their drinking water, following the manufacturer's instructions.

Flaxseed is rich in omega-3 fatty acids, which improve heart health (Sextro, 2019). This supplement is incorporated into chicken eggs. Excess flaxseed may harm chickens, so use the actual seeds and not flaxseed meal.

Kelp is nutrient-rich and includes folate, calcium, and vitamin K, which promotes healthy bones (Sextro, 2019). Add it to feed or include it in homemade treats.

Sunflower seeds boost the cardiovascular system, reduce inflammation, and help produce the oils used for preening. Seeds without shells are best, but if you use ones with shells, soak them in water overnight to soften them before feeding them to your chickens.

Spice Up Your Birds' Lives

Besides flavoring foods, many spices are used for human and animal health. Give certain spices to your chickens in moderation to improve their health and well-being. Spices are often very strong, so add just a sprinkling to your chickens' feed. Chickens don't have a broad taste spectrum and lack "heat" receptors, so you can give them cayenne pepper with no ill effects.

Tip: Adding a sprinkling of cayenne pepper and dried mint leaves to your feed bins will help keep lice out of the feed.

Below are a few common spices, together with their benefits:

- Black pepper contains vitamins and other nutrients. It is an excellent digestive, is an anti-inflammatory, and has antibacterial properties. It contains antioxidants that help flush toxins out of the body. Black pepper is helpful for coughs and benefits respiratory health.
- Cayenne pepper added to feed in winter combats the cold, raises circulation, and prevents frostbite. It will help layers maintain egg production. Cayenne pepper boosts the immune system and makes lovely orange-yellow egg yolks.
- Cinnamon reduces inflammation and infection, is antibacterial, and has antioxidant properties. It guards against neurological diseases and relieves congestion and coughs. Cinnamon thins the blood,

thereby improving the circulation and blood flow to wattles, legs, and feet, helping to prevent frostbite. Chickens have complex respiratory systems, so giving them cinnamon will keep them clear and functioning properly.
- Ginger is an excellent supplement for laying hens. Add 0.035 oz of ginger per 2.2. pounds of chicken feed (Steele, 2015). Ginger improves egg laying, also resulting in bigger eggs. It is an antioxidant.
- Turmeric provides the yellow color in curries but also has excellent health benefits for your flock. It is a powerful anti-inflammatory that can relieve the swelling caused by bumblefoot. In addition, it is antibacterial, antiviral, and an antioxidant. It improves digestion and maintains healthy skin, eyes, and brain function. Turmeric relieves wry necks in chickens. Applied topically, it speeds up healing and repairs damaged skin. Give 1/3 to 1/2 teaspoon in the feed to a sick chicken (Steele, 2015).

WHAT *NOT* TO FEED CHICKENS

There are some things that are toxic to chickens—and birds in general—and these should be avoided. Here's a list of foods to avoid:

- The pit and peel of avocados contain persin, which is very toxic to birds, including chickens. You can give them the flesh, however.

- If you have dogs, you'll know that chocolate is toxic to them, but it's just as poisonous for poultry. So, no chocolate treats for them.
- Citrus isn't the birds' favorite, and it's not recommended for them, either.
- Apple seeds contain small amounts of cyanide, which is a poison, so don't feed them to your chickens.
- Raw potatoes and green potato skins are poisonous to chickens.
- Dry beans and undercooked beans contain a compound that inhibits digestion in birds, so your flock will become undernourished. Not a good idea.
- Junk food isn't good for either you or your birds, so don't give it to them.
- Rhubarb has a laxative effect. If the plant is damaged by severe frost and cold, it produces oxalic acid, which is toxic to chickens. No rhubarb pie leftovers for them.
- Avoid feeding your chickens salty foods, as salt can build up over time and affect egg quality. It can dehydrate your hens.
- Don't give your flock moldy, rotten foods as these make their poop wet and may contain toxins that will compromise their health.
- It is not recommended that you feed chickens coffee grounds or soft drinks.

* * *

If you are enjoying this book, please consider leaving a review.

* * *

SUMMARY

Feeding your chickens correctly is imperative to ensure that they are getting all the nutrients they need for good health, as well as maintaining egg production or providing tasty, tender meat. Buy the best feed you can afford. Free-range chickens will benefit from foraging insects, small animals, greens, weeds —and even your vegetables and flowers!

Egg-laying chickens usually need a calcium supplement in the form of crushed oyster shells or even dried, crushed eggshells in a pinch but beware of using too much of the latter as they might start eating their own eggs. Chickens eat almost everything, but by far, the bulk of their food should be a properly balanced feed with only a small percentage of their diet consisting of treats such as table scraps, scratch grains, and mealworms. Certain foods are toxic to poultry, so avoid giving them to your birds.

Chickens, in general, benefit from vitamins and other supplements added to their feed and water or provided separately for them to consume voluntarily as needed. There are several natural and commercial supplements available. Spices and herbs such as oregano can also help to maintain flock health and prevent certain diseases.

In the next chapter, you'll find out how to achieve your chicken-raising goals, whether you are raising chickens for eggs, meat, or as show chickens.

CHAPTER 5
Tips for Reaching Your Chicken-Raising Goals

There are more chickens in the world than any other kind of bird.

— RACHEL SMITH

As mentioned previously, everyone's goals for backyard chicken keeping are different. You might be raising them for eggs, meat, or even to exhibit rare and beautiful breeds. This chapter will provide you with some hints and tips on how to get the most out of your chickens, as well as scheduling a program for their care.

TIPS FOR RAISING EGG-LAYING CHICKENS

There are times when your chickens will lay fewer eggs, but there are several things you can do to ensure that their egg production improves and remains constant:

- As mentioned in the previous chapter, give them sufficient good-quality food. A balanced diet for layers should include protein from sources like wheatgrass, barley, kelp, and alfalfa in addition to a good layer feed, as well as beneficial herbs, kitchen scraps, garden weeds, mealworms, and some fruit.
- Give them supplementary calcium by supplying ground oyster shells. In a pinch, you can also dry chicken eggshells, grind them up, and add them to food.
- Ensure that they have enough water and that it's always clean. Use waterers to keep the water clean and ensure a constant supply.
- Provide a well-ventilated chicken coop that is always clean.
- Check your chickens regularly for parasites, especially mites.
- Make sure your chickens are adequately protected from predators, and shut them in the coop at night. This will prevent stress, which can compromise egg production.
- If possible, let them free-range. Pastured chickens are healthier, happier, and have access to more nutritious foods. If you can't free-range your chickens, at least provide them with a decent-sized run.
- Some chicken keepers have found that their hens lay more eggs if there is a rooster in the flock.

Collecting Eggs

Collect fresh eggs early in the day or as soon as they've been laid. Do this two or three times a day, if practical, to ensure that the hens don't eat them or become broody, that eggs are not laid on top of one another, and that they are clean (Arcuri, 2023). The chickens might poop on the eggs or sleep on them, which can damage them if they are left in nest boxes overnight.

There needs to be plenty of wood shavings or straw in the boxes—the lining should be at least 2 inches deep (Manitoba Agriculture, n.d.). If an egg breaks and soils the lining, clean it up quickly and replace the shavings or straw. Keeping the boxes clean and fresh encourages the hens to lay there and not elsewhere.

Cleaning Eggs

Use either a dry or wet method. The dry one is preferable because it maintains the protective antibacterial layer on the egg. Wipe the egg gently with an abrasive sponge or loofah to clean dirt and poop off the shell.

If the eggs are particularly dirty or have yolk stuck to the shells, you might need to use water. Carefully wash the eggs under running water that is warmer than the eggs but not hot or hairline cracks may form. Wipe the egg with a paper towel and place it in a wire rack or clean, open carton. Spray them with a diluted bleach solution.

Storing Eggs

Once they are clean and dry, pack the eggs into a carton and label it with the collection date. Store the eggs in the refrigerator if they have been washed with water. You can store dry-washed eggs at room temperature, although eggs last as long as

4–5 weeks if refrigerated (Arcuri, 2023). They should be baked or hard-boiled if stored for longer. Wash dry-cleaned eggs before cooking.

If an egg floats in a bowl of water, the air space inside has evaporated, and it is probably inedible. Compost these eggs.

Dealing With Cracked Eggs

If you are finding numerous cracked eggs in your boxes, there's a good chance your hens are responsible. Old hens, especially if they are over 60 weeks old, lay eggs with thinner shells that are more likely to crack (Manitoba Agriculture, n.d.). Extra-large and jumbo eggs are more inclined to crack than smaller ones. If the hen stands when laying as opposed to squatting—most hens squat to lay—then the eggs could crack. If the chickens are disturbed, especially at night, they may lay eggs with weaker shells. Feed or nutritional problems often cause poor shell quality.

Gather the eggs slowly and place them in egg trays rather than piling them in egg baskets. Put the eggs' narrow end down in the trays. Keep jumbo eggs separate, and don't stack trays more than six high (Manitoba Agriculture, n.d.). Always handle eggs carefully.

TIPS FOR RAISING MEAT CHICKENS

This is one of the quickest and most rewarding aspects of backyard chicken keeping. A chicken can transition from chick to freezer in as little as 6–12 weeks, depending on the breed and desired slaughter weight (Hamre & Phillips, 2021).

Check beforehand whether your local ordinances allow you to slaughter and process chickens in addition to raising them. If not, you will need to find a processor. Some states require chickens to be processed in a certified facility, so find out what the requirements are before you invest in meat chickens. Ensure that you have sufficient freezer space if you intend to eat the chickens yourself.

Small flocks usually only produce sufficient meat to feed a household. If you want to sell your poultry, consider local market prices for different classes of meat rather than retail store prices.

Meat classes for chickens are:

- Capons—male birds slaughtered at 3–4 weeks and marketed at 18 weeks
- Cornish game hens slaughtered at 5 weeks of age
- broilers or fryers slaughtered at 7–9 weeks of age when they weigh 3–5 pounds; the dressed carcass weighs around 2 1/2 –4 pounds
- roasters grown for 12 weeks or more (Hamre & Phillips, 2021)

Most potential customers want heavier chickens than those available at stores.

Slaughtering chickens at different times will give you several different weights. You could slaughter a third of your flock at 5, 7, and 9 weeks of age, for instance (Jacob, n.d.b).

Here's how to work out your cost per pound if you want to sell meat chickens:

1. Calculate the cost of the chicks—add a few extras to allow for any deaths.
2. Work out your feed costs. Chicks need about five pounds of feed for the first six weeks and 8–9 pounds until eight weeks for commercial breeds (Hamre & Phillips, 2021). Consider what the final use of the chickens will be: chickens for roasting and capons will need more feed per pound of meat than fryers.
3. Don't keep birds after they have reached the desired weight.
4. Factor in depreciation—over 10 years for equipment and 20 years for housing (Hamre & Phillips, 2021).
5. Estimate the costs of things like bedding, heat for brooding, lights, and other sundries. Factor in labor charges for cleaning, help with chicken care, and so on.

Work out your costs per pound of meat produced. Divide the total cost per bird by the anticipated pound of market weight. The cooking weight will be 70–75% of the live bird weight (Hamre & Phillips, 2021).

Ordering Chicks for Meat Chickens

Consider the slaughter date when ordering chicks. Cornish cross broilers, for example, need only 6–8 weeks to reach a marketable carcass of 4–6 pounds. Other breeds may take up to 12 weeks (Hamre & Phillips, 2021).

Order cockerels, pullets, or a straight run (mixed batch). Cockerels cost more but are usually about one pound heavier than females at processing (Hamre & Phillips, 2021). Ask the hatchery to vaccinate the birds against coccidiosis.

Requirements for Meat Chickens

- Coops should be quiet and draft-free. Allow 1.5 square feet of space per chicken (Jacob, n.d.b). They may require a heat lamp for warmth.
- Allow for expansion. Meat chickens grow rapidly—they can double in size in just a few days and grow rapidly, especially in the first six weeks (Jacob, n.d.b). They need a regular supply of clean, fresh water. One-quart waterers will be sufficient at first, but you'll later need to expand to one-gallon waterers (Jacob, n.d.b). The same applies to feeders—you can start off with smaller ones, but you will need to expand them as the chickens grow.
- Broilers need ample bedding to keep warm and absorb moisture. Cover the floor with wood shavings, sawdust, or rice hulls to a depth of 3–4 inches (Jacob, n.d.b). Remove any caked or matted litter daily. The remaining litter should be turned over or stirred up so it absorbs more moisture and lasts longer between changes. Change it weekly. Never use newspaper, cardboard, or plastic, as these become slick when wet, and you don't want your chickens to develop leg problems.
- Broilers need outdoor space, as this prevents aggressive behavior. Ensure that the run is large enough and has shaded areas, together with sufficient foliage cover.

TIPS FOR RAISING SHOW CHICKENS

Hybrid or mixed-breed birds are unsuitable for shows and exhibitions. You will need to raise purebred chickens, and your entry should comply with the definitions of the American Poultry Association's American Standards of Perfection. Familiarize yourself with them so that when you are ready to enter a show, you know which birds to select.

Sourcing Show Chickens

Avoid buying chicks from feed stores or farmer's supply outlets where they are selected for quantity rather than quality. Find show chickens on the internet or at your local poultry fanciers club, where breeders might have ones you want. The Livestock Conservancy will help you find heritage chickens. Be prepared to pay for your show chicks, however, as these more spectacular breeds come with a higher price tag.

Don't succumb to chicken math—only buy the number of chickens you are able to house, rear, and nurture. Experienced farmers usually have just 1–5 varieties. Those new to showing chickens should stick to just one or two breeds to start off with. Having said that, purchase and raise several birds for the class you intend to show so you will have a few to choose from as show time approaches. This means you have standby chickens should your preferred one decide to start molting a few days before.

Requirements for Rearing Show Chickens

- Don't let your breeds intermingle—crossbreeds can stop your chicken show dreams before they even get started.
- Don't let your show chickens free-range as they could be injured while outside.
- Don't let your show chickens socialize. This is particularly difficult as chickens are sociable birds—but your prize chicken could end up with tatty feathers after being accosted by other hens or your rooster.
- House your show chickens in special enclosures that accommodate three birds at most. These allow enough room to move and sufficient companionship to combat loneliness. Use rabbit hutches, portable dog crates, or chicken tractors to house your show birds.
- As the show date draws nearer, select only the best—birds as close to perfect as you can get them.

Lessons From Experienced Exhibitors

- You must know the sex of the bird you are exhibiting—the organizers often get it wrong.
- Securely close cage doors and keep a net handy in case your prize chicken escapes.
- Don't get upset if the judge seems young and inexperienced—stand-ins happen.
- Label your belongings.
- Be courteous towards your fellow exhibitors.

Caring for Show Chickens

Always provide plenty of fresh, clean water so chickens don't become dehydrated. This ensures that they grow and show well. Provide them with high-quality feed, as they won't be able to go out and forage like your regular flock.

Handle the birds constantly, starting as soon as they arrive. This will get them used to being handled at shows and exhibitions. Before a show, put them into a cage similar to the one that will be used there and take them in and out to get them used to being moved around.

A few days before the show, examine the birds for physical defects. These include cuts and tears, broken or disjointed bones, skin or flesh bruises (apart from on the wing tips), breast blisters, insect bites, and external parasites. Extremely dirty birds should be excluded. Give your chickens a bath a few days beforehand so the natural feather oils return after the wash, ensuring that their feathers look fabulous.

It's Show Time!

Transport the birds in large, roomy cardboard boxes. Line the boxes with litter to protect the birds against bruising. Don't bruise the birds while transporting them, and never drop the container.

At the show, keep the cage clean so your bird doesn't become dirty. If this happens, wipe down any unclean areas before it goes before the judge.

Apply comb reddening an hour before judging, as this improves the overall look of the bird.

YOUR CHICKEN CHORES

Daily Chores

- Open your chicken coop in the morning. If you have automatic doors, the timer will open them at the pre-set time. Take along your bucket of kitchen scraps, together with the day's feed.
- Clean water bowls and containers with soap and water as needed. You can use a little bleach to sanitize the container but make sure you rinse it thoroughly. Refresh the water and make sure it's clean. Chickens won't drink dirty water, and it can be detrimental to their health.
- Feed the chickens. You can either free-feed them or give them a set amount of feed daily.
- Collect eggs from the nest boxes.
- Remove any dirty bedding. Add clean bedding if needed.
- Add clean bedding to the nest boxes if required.
- Clean overnight poop from beneath the roosting poles—this will keep the coop clean and fresh-smelling. (This is always recommended – but I don't know anyone that actually does this every day. Use your judgement about a cleaning schedule.)
- Assess your chickens. Make sure they are bright-eyed, alert, and healthy.
- Provide treats at a specified time during the day.
- Bring the chickens in at dusk and close up the run. Call any that aren't there and ensure that they enter the coop before you close up for the night. If you use

an automatic timer, it will close the doors at the pre-set time. Make sure all doors and windows are closed, secured, and properly latched.
- Store the feed bucket.

Weekly Chores

- Inspect your chickens thoroughly to ensure that they are parasite-free, uninjured, healthy, and disease-free.
- Remove the bedding completely and replace it with clean bedding. Doing this weekly helps control parasites and moisture, which is where harmful bacteria could breed. It also prevents frostbite in winter.
- Change nest box bedding.
- Rake the run to free it from chicken waste and any dirty litter or bedding material like straw. Keep the run floor clean to maintain optimal chicken health.
- Check fences, irrigation equipment, and any evidence of predators.
- Manage rodents.

Monthly Chores

- Buy feed at the feed store and store it securely in rodent-proof bins or containers.
- Clean the chicken coop thoroughly and disinfect it if you are experiencing parasite problems. While you are at it, change the bedding too.
- Clean the outside run and level the floor.
- Undertake repairs and maintenance of the coop.

Twice-Annual Chores

- Scrub out the coop with soap (ordinary dish soap is fine), water, and a scrubbing brush. Apply diatomaceous earth to nooks and crannies to prevent insects from inhabiting them.
- Freshen up the dust bath with peat moss, wood ash, and sand.
- De-worm your chickens in spring and fall.
- In the fall, prepare your chickens for winter. Get out your water heaters and decide if you need lights to keep your hens laying. Make sure there is enough roosting space for all the chickens to keep warm. Don't heat the coop.
- In spring, assess your flock. Are you going to buy more chicks or adult chickens? Do you need to enlarge the coop or build another one? Do you need more outside runs?

Annual Chores

- Repair the coop if you haven't done it previously that year. Pay special attention to doors, locks, hardware cloth, and any gaps in the planking.

Chicken keeping is not difficult, but you need to be on top of your chores to ensure productive, healthy, and happy birds or beautiful show chickens.

SUMMARY

Whether you are raising chickens for eggs, meat, or exhibitions and shows, there's a lot you can do to maximize your yield or ensure your birds receive accolades at shows. Your chickens must have secure, adequate housing, plenty of water, the right type of food, and be safe from predators. This will reduce stress and improve egg production, the quality of your meat, and the attractiveness of your birds, depending on your chicken-raising goals.

Your chicken chores include daily, weekly, monthly, biannual, and annual chores, ranging from day-to-day care to giving the coop a thorough clean-up and your runs a good going-over. Ensuring that you give your chickens the best possible care allows you to fulfill your chicken goals.

Chickens are very social, interactive birds. In the next chapter, you will find out all about the "language" of chickens, their pecking order, and how they react to one another and to humans. This will enable you to better understand your chickens and the interactions you might observe while working with them.

CHAPTER 6
Chicken Language

I dream of a better tomorrow where chickens can cross the road and not be questioned about their motives.

— RALPH WALDO EMERSON

Chickens are sociable, interactive birds and have a surprisingly complex society. If you observe them for a while, you'll notice that they have a definite "pecking order" or hierarchical structure. By watching them, you'll soon discover whether they are comfortable and content or unhappy. Chickens can be aggressive, and you might need to discipline them so they know that you're not only their caregiver but you're also the boss.

This chapter aims to give you a window into the world of chicken behavior, enabling you to better understand your birds and what makes them unique.

THE ORDER OF THE CHICKEN

The black hen sat atop the nest box area inside the coop. She refused to come down. When I asked the chicken keeper what the problem was, she said, "That hen is afraid of the rooster. I feed her separately so she can feel more comfortable" (Langston, A., personal communication, December 27, 2022).

Chickens are not democratic or altruistic. They firmly believe in the principles of "might is right" and "survival of the fittest." They aren't too worried about elbowing other chickens out of the way at feeding time, competing for the best spaces at the dust bath, or grabbing the sunniest spots in the chicken run on a chilly day.

But if you watch them closely, as Norwegian zoologist Torleif Ebbe did after he was relegated to look after his family's flock in 1904, you'll notice that there's a definite hierarchical structure in chicken society (Moore, 2018). Like other animal societies, some flock members are more dominant than others. If other chickens transgress their boundaries, the more dominant ones give them a painful peck, hence the term "pecking order" as it relates to chicken hierarchies.

Purpose of the Pecking Order

The pecking order literally determines in what order chickens access resources like food, water, and dust-bath areas, not to mention the comfiest nest boxes and the best places on the roosting bars. The pecking order naturally establishes early, and the birds live in relative harmony when they are hatched and raised together. The pecking order applies to all chickens in the flock, not only the strongest ones who have leadership roles. Every chicken knows her place and isn't shy about asserting it if another threatens to jump the queue at the feeder or usurps a better nest box.

At the top of the pecking order are the alphas—the strongest, healthiest birds. They guard the flock, keeping them safe from predators and ushering them to safety should one appear on the ground or in the sky above the coop. Alpha chickens also find food delicacies. Surprisingly, they rarely pull rank at the feeder, allowing the others to have their fill while they keep an eye out for predators before eating.

It's not only gender that determines dominance: Old, savvy hens can take the top spot in the chicken flock hierarchy, especially in hen-only flocks. There is some scientific evidence that dominance might be inherited rather than learned—dominant chickens are often the offspring of dominant adults.

In mixed-gender flocks, both roosters and hens compete for the alpha position. A beta rooster might share the top spot but isn't allowed to challenge the alpha and will be severely reprimanded if he tries.

When the alpha rooster is past his prime, he will be replaced—he might surrender his position in a pitched battle or quietly retire. When the new incumbent takes the reins, the entire flock's pecking order needs to restart.

A single rooster can manage a flock of 10–15 hens. If the flock comprises over 30 birds, the roosters usually divide up the chickens between themselves, each looking after around 10 birds (Moore, 2018). In a hen-only flock, one of the hens becomes the alpha, with the others descending through the pecking order.

Gender and the Pecking Order

If you have a rooster, chances are he'll be the top bird in your flock's pecking order. In a female-only flock, the strongest hen assumes the alpha role. While she also governs the coop with an iron hand, the interactions are calmer than when a rooster is present.

In a mixed-gender flock, there are three social orders operating simultaneously to ensure the viability of the flock—in the wild, a flock is only as strong as its weakest members:

1. rooster to rooster
2. hen to hen
3. roosters to hens (Moore, 2018).

Roosters copulate with the hens continuously, and a favorite hen can end up with a sore back, missing tail feathers, and other injuries that could lead to health problems. If the rooster-to-hen ratio is too low, then the coop dynamics quickly become very volatile indeed.

Chickens will do their utmost to maintain the pecking order. They will kill other chickens, as they are naturally cannibalistic, and might band together to inflict serious injuries on an errant chicken. These extremes are unlikely to occur if your flock is happy, healthy, has sufficient space, and isn't bored.

MANAGING THE PECKING ORDER

You, as the chicken keeper, are responsible for managing your flock's pecking order and making sure the alpha roosters and hens—or any other birds—don't misbehave.

Roosters

Having the correct number of roosters is essential so your hens aren't exhausted from the alpha male's constant sexual advances or aggression. Aim to have 1 rooster per 10–12 hens. If you have six hens or less, it's not advisable to get a rooster at all (Barth, 2016). Don't have too many roosters as they will fight among themselves.

Flock Management and Coop Design

Your chickens must have plenty of space—this makes for tranquility, and pecking order disputes will be less likely to erupt. Allow 1–4 square feet of space per bird indoors and 4–8 square feet per bird in the run (Moore, 2018; Barth, 2016). If you have roosters, double those figures.

Coop design is important. Hang feeders and waterers in the middle of the indoor space or in an open area rather than in a corner, so birds can move around them without crowding. If you have several birds, place more feeders and waterers inside and outside the coop so birds don't all congregate in one spot.

Ensure that there are sufficient nest boxes and plenty of roosting space, as recommended previously.

If your chickens are confined for long periods, use boredom-busters like hanging pecker blocks or fresh greens. Change these frequently to keep them entertained.

Birds may become aggressive or bored if there is too much light or if lights are on for long periods. This may add to pecking order disputes. Even in the brooder, chicks will start pecking one another if the lights are always on. Limit illumination to 16 hours per day at most (Moore, 2018). Use an alternative heat source to enable chicks to develop regular biorhythms.

Mitigating Territorial Disputes

Warfare is likely to break out in your coop when new chickens are introduced. If your existing flock outnumbers the younger chickens, then there could be trouble—chickens tend to gang up on a bird that is wounded and bleeding. If the wounded chicken isn't separated from the others, it could be pecked to death and even eaten.

To avoid these unpleasant skirmishes, introduce new birds slowly. Fence off an area inside the chicken run for newcomers, so everyone can have their say without the new kids on the block being victimized. Remove the fence after a few days but keep a sharp eye on the new chickens. If any are wounded, take them out immediately. Try reintroducing them a second time after you have treated their injuries.

Chickens recognize one another, with hens able to identify as many as 30 other individuals. Large flocks of over 60 birds are more tolerant, and the intense interactions associated with the pecking order largely diminish (Jacob, n.d.a).

Humans in the Pecking Order

In case you're wondering, you—and your family—are also part of the pecking order. Hens automatically respect you as another type of alpha, but roosters might decide you're their competition and need to be reminded of your place.

Aggressive roosters can be annoying. But they can also be dangerous, especially around small children. Put on thick rubber gloves and pin the culprit to the ground whenever he comes for you. This establishes the pecking order and often sorts out the problem. If not, you will need to either rehome the rooster or slaughter him for Sunday lunch.

COMMON CHICKEN BEHAVIORS

Foraging and Feeding

Chickens spend 61% of their active time on foraging and feeding behaviors, including scratching and pecking at the ground to find food, besides actually eating (McCrea & Baker, 2022). If a chicken sees another one feeding, it wants to do the same. This social behavior is known as contra-freeloading and has nothing to do with food availability. This helps chickens find food and monitor predators simultaneously. Chickens sometimes feed in groups, with one group eating while the other watches for predators. Chickens must forage, or they will become frustrated, exhibiting undesirable behaviors like excessive pecking, egg eating, and cannibalism.

Resting and Roosting

Chickens need to rest and sleep at night, and they start roosting at dusk. Roosting on perches elevates the birds off the ground, protecting them from predators. Subordinate birds can also get away from more dominant ones for a while. Roosting improves feather condition, foot health, and bone strength. Chicks begin roosting when they are one or two weeks old (McCrea & Baker, 2022).

Comfort Behaviors

Comfort behaviors usually involve the chicken maintaining and caring for its body and plumage, as well as stretching. For example:

- Preening is when a chicken runs its beak over its feathers to knit them together. This improves waterproofing and insulation. While preening, the chicken spreads oil from the preen gland over its feathers, keeping them glossy and water-resistant. It's a social activity done daily by the entire flock.
- Dust bathing is when a chicken finds a patch of loose sand and lies down in the dirt, kicking the sand over itself. This manages external parasites by loosening their hold on the skin. It also conditions the chicken's feathers, getting rid of dead skin and old preen oil. Chickens dust bathe in groups, and hens actively look for suitable dust baths if one is not provided.

Preening and dust bathing are motivational behaviors. Chickens must be allowed to perform these activities, or they will become frustrated and behave badly.

Exploratory Behavior

From the time they are day-old chicks, chickens explore their world. They peck at objects and scratch the floor. Their beaks have touch receptors that provide sensory information to the birds. By pecking, the birds learn about different objects and their environment.

Social Behaviors

Chickens learn behaviors by watching other flock members. They also watch humans and learn from us too.

Social Learning

When a new bird enters the flock, the other birds watch how another, more dominant bird reacts to establish where the newcomer should be in the pecking order. If the dominant bird loses, then the other birds know it is a higher-ranking bird and won't challenge it. But, if the reverse happens, then the observing bird will challenge it to establish the basis of their relationship.

When they are chicks, birds watch their mothers to find out how to find food, what to eat, how to perch, and where their home is. They also watch other chicks.

Communication

Chickens communicate through displays and vocalizations if flock members aren't nearby. Displays are avian body language. Birds change their posture or the position of certain parts of their body, like the head, tail, or feathers. They indicate personal space preferences, health, or flock hierarchies and are also used during territorial disputes and courtship.

While most of us are familiar with the rooster's loud crow or the soft clucking of hens in the coop, chickens actually make as many as 30 different sounds to communicate with one another (McCrae & Beacon, 2022). For example, they have two different alarm calls—one for aerial predators and one for terrestrial threats.

Chicks even vocalize in the shell, "talking" to their mothers before they hatch and communicating with other chicks, so hatching is coordinated.

Chick-Specific Behavior

Chicks instinctively preen, scratch on the ground, fear stinging insects, and try to catch flies. Newly hatched chicks seek warmth automatically. They pursue the hen, especially for the first eight days after hatching (Jadob, n.d.a). Chicks can identify their mother's sounds and find her by sight, although the latter isn't always as effective.

Other behaviors are learned, including drinking water and avoiding their own poop. Chicks learn by imprinting, following the first moving object they see after hatching, and identifying it as their mother. Their mother teaches them everything they need to know to survive and stay safe. If she isn't available, they will imprint on other chicks.

Chicks initially tolerate one another, but after 16 days, they begin fighting to determine the pecking order (Jacob, n.d.a). This is usually established by the eighth or tenth week if the flock consists only of females or is a small one. Groups of male birds can take several weeks to establish the pecking order.

DISCIPLINING YOUR CHICKENS

Disciplining a chicken is very different from disciplining other household pets. But it can be done. Before attempting this, however, it's important to observe your chickens, become part of the flock in a sense, and get some insight into their pecking order.

Stopping Chickens From Pecking You

To your chickens, you're the alpha, but some might decide to challenge you. They do this in the same way they would with another hen, pecking your hands, legs, or back when you're in the coop or run. It's important to stop this behavior immediately, or it will persist.

Grab the errant hen, lift her upside down by her feet, and hold her in that position for a bit. When she calms down, put her back on her feet. If she does it again, repeat. The second time, she'll know you're the boss.

Alternatively, use a spray bottle or water gun, but then she won't perceive you as alpha. Show her dominance, and she'll respect you.

Don't wear any jewelry, especially on or close to your face, including earrings, when working with your chickens. The bling will encourage them to peck your face. This is important if you like to "cuddle" your chickens.

Taming an Aggressive Rooster

An angry hen is one thing, but an angry rooster is an entirely different bird. Roosters naturally fight to defend and maintain their territories, and they'll claim the coop or run as their own if kept with other chickens. Roosters may attack you if they perceive you as a threat. The good news is that you can discipline roosters without having to hurt them, and they'll learn the lesson for keeps.

First, adopt a bigger posture by standing straight with your head high and your arms out at your sides. This will dissuade him from attacking you. Carry a stick to block potential attacks and walk determinedly towards him. Make a noise and wave your arms as though you are going to hit him. Never run away or back down, even if he approaches you, as he'll believe he's won and you are not actually Alpha One.

Grab him carefully but firmly, avoiding his spurs and beak. Pick him up and hold him for a while. He will try and get away, which is why you need to be firm. After holding him for a while, he should get the message.

If he continues trying to attack you after you have put him down, try and pin him to the ground until he settles down. This will finally show him that you are Alpha One and you'll have won the dominance war.

There are, however, roosters who never give up. As previously mentioned, aggressive roosters can pose a threat, and it's best to rehome or cull them if their negative behavior continues or worsens.

Stopping Aggressive Behavior in Your Flock

Bullying is different from maintaining the pecking order—it's constantly harassing another chicken to intimidate or harm her. If a chicken becomes overly aggressive, spray her with water or make a loud noise. Do this immediately and be consistent because chickens learn by repetition. If you do this continuously, she'll eventually stop.

There are a few things that make chickens behave aggressively:

- Chickens hate change. A change in the environment —like introducing new chickens or changing something in the coop—can trigger unexpected aggression.
- If the aggressor feels threatened by predators or anything else, she may take it out on a weaker chicken.
- Overcrowding can cause aggressive behavior, so make sure there is room for everyone. The flock might need more outside time too.
- Chickens are curious birds and like to explore. If they are cooped up with little to relieve the boredom, then they might become aggressive. Relieve winter boredom by supplying a cabbage tetherball, flock blocks, handfuls of scratch, or more table scraps. Thin out the flock or extend the amount of room available by adding an apron to your coop.
- Pain and anxiety may also trigger aggression. If a hen goes off her food unexpectedly, watch her behavior. If you suspect that she is sick or injured, treat her or take her to the vet.

Sometimes chickens will attack an injured bird. They want the sick chicken away from the flock, especially if they believe the chicken has a transmittable disease (often she doesn't, but this is a survival instinct.) If the chicken has an open wound, they will peck it continuously until the bird is dead.

Remove the wounded, injured, or sick chicken. Put her in an enclosed, secure area. Keep her close to the flock but separated by a mesh fence so she can easily be reintroduced later. Check your chickens regularly for wounds or injuries—they might hide them to avoid being bullied.

Dealing With Bullying

There are several methods you can use to control an unruly, antisocial hen:

- A well-timed squirt with a water pistol will stop the aggressor in her tracks. You will need to do this repeatedly, however, and it may be time-consuming. You'll also need to spend a lot of time observing your chickens so you can hone in whenever the bully strikes.
- Try using a "pebble can." Fill an old tin can with pebbles and tape it up securely. Whenever you see bullying, shake the can vigorously. The noise will stop the offending chickens.
- If that still has no effect, try a product called "pinless peepers." These are similar to sunglasses but don't allow the chicken to see in front of them. The chicken can continue with her regular activities—except peck at other chickens' feathers.
- As a last resort, send the bully to chicken jail. Put her in a separate cage away from the other hens but where they can still see her. Most chickens

rehabilitate within 3–7 days, but others take longer ("Chicken Bullying," 2021). The pecking order will need to readjust when you return her to the flock.
- Alternatively, split the flock into two groups, forcing the pecking order to re-establish itself. Put two sets of feeders and waterers in the coop so any excluded birds are still able to feed and get their nutrients (Smith, 2020a).

Disciplining Chickens That Destroy Your Yard

You cannot discipline your chickens to prevent them from scratching, dust bathing, eating new seedlings in your vegetable garden, or for eating your flowers. This is natural chicken behavior.

Prevent your chickens from accessing your vegetable garden or flower beds by fencing these areas off with a high mesh fence. This will prevent parts of your garden from being decimated by your birds. Provide your chickens with an area where they can dust bath—sometimes, they target your vegetable garden just so they can dust bath.

SUMMARY

Chickens are highly sociable birds that learn from one another and benefit from doing things together, like foraging and feeding, roosting, preening, and dust bathing. If they aren't allowed to do these things, they may become aggressive. They are curious creatures that love exploring—they do this from the earliest days of chick-hood and never really stop. They communicate by displaying their bodies in certain poses and postures and also through a surprising number of vocalizations.

There may come a time when you need to discipline your chickens to stop them from pecking or attacking you—or one another. There are various methods that can be used to reinforce the fact that you are alpha or to prevent your birds from injuring one another as a result of stress, boredom, or ill health.

In the next chapter, you'll find out how to tell if any of your chickens are unwell or if their health is below par, as well as how to remedy minor chicken ailments.

CHAPTER 7
What's Wrong With My Chickens?

I did not become a vegetarian for my health, I did it for the health of the chickens.

— ISAAC BASHEVIS SINGER

If you look after your chickens properly, giving them nutritious food, plenty of water, and clean, sanitary coops, runs, and nest boxes, then they shouldn't have serious health issues. There are some common ailments you might see in your chickens, such as cuts or peck marks, foot injuries, and eye problems that you can solve yourself. More serious diseases and infections will require veterinary intervention.

COMMON HEALTH PROBLEMS

Egg Laying Difficulties

Egg-laying difficulties are among the most common problems chickens experience. Observe your hens to establish the cause. Parasites, infections, and stress are just some of the reasons why your hens may abruptly stop laying. Symptoms may include loss of appetite, lethargy, abnormal poop, weakness, and respiratory problems.

Egg-laying issues can range from soft-shelled eggs to egg binding (when the egg gets stuck inside the chicken.) In general, vitamin supplements and oyster shells added to the feed will resolve most problems. For egg binding, try giving the hen a warm bath followed by an application of Vaseline inside and around the vent to help her pass the egg. Place her in a quiet, secluded area away from the other hens. If she still can't lay the egg and is distressed, contact your vet urgently.

I once had a very good layer stop laying eggs for two days, which was unusual, and I noticed a bump. I thought she was eggbound and called a vet. As it turned out, she was taking a break from egg laying and had a full stomach—the lump was a FOOB or food boob. When this happens, the front of the chicken may looks like they have an open zipper. This is actually their crop, sometimes referred to as a "broody packet" or "food boob." The chicken's down emerges from beneath the thicker, bigger feathers, which form a shield around them. Hens sometimes fluff the down feathers outwards underneath to keep the eggs warm.

Cuts or Peck Marks

Hens are feisty, and your birds will inevitably have cuts, peck marks, and other minor wounds, often inflicted by their companions. Cuts and peck marks are obvious—and chickens may have bald spots, missing feathers, scabs, or cuts on places like their backs.

Remedy excessive pecking by doing things like enlarging the coop or isolating the aggressive birds. Treat abrasions and cuts with colored wound spray. This both treats the sore and hides it from the other birds so they don't keep pecking at it.

Foot Injuries

While foot injuries aren't usually serious, they can be challenging. If a bird doesn't want to put weight on one of its feet, the foot might be injured. Some injuries are simply cuts or entanglements, while others, like bumblefoot, are caused by infections. Chickens may be lethargic and reluctant to move around. Pus-filled abscesses under the foot are indicative of bumblefoot.

It is usually sufficient to treat the foot with antiseptic wound wash, after which it should be bandaged to prevent infection. Isolate injured birds so they can heal. Treat bumblefoot with antiseptic wound wash, an antibiotic cream, and gauze. If this doesn't resolve it, a vet might need to drain the abscesses.

Eye Issues

Healthy chicken eyes should be clear and bright, wide open, have a definite black pupil, and have no discharges or swelling. Dirt, abrasions, wounds, chemicals, and many other things can cause eye problems in chickens. The eye may become cloudy, and you might wonder whether it can be saved. Clean it with saline solution—the chicken may not want to open the eye due to photosensitivity—and apply an over-the-counter medicated eye gel. This frequently sorts out the problem.

If the eye is wounded and bleeding, apply pressure with gauze until the bleeding stops. Treat it with an antibacterial wound spray and cover it with gauze if necessary. Alternatively, coat the eye with blue antiseptic to prevent the other chickens from pecking at it.

DISEASES

Parasitic Diseases

Mice, lice, ticks, and worms cause parasitic diseases in chickens. One of the primary causes is poor hygiene, especially if coops are not cleaned frequently and are filled with soiled bedding. Second-hand coops may also come with parasites.

Signs that your chickens are infected with parasites include listlessness, anemia (pale comb and wattles), itchiness, irritation, restlessness, reduced egg production, weight loss, matted feathers, feather loss, and reluctance to enter the coop.

Chicken Lice

Chicken lice lodge around the vent area and under the wings. They may survive for several months. It is difficult to remove lice eggs without removing the bird's feathers. Infestation is more likely in autumn and winter. Lice are spread by contact between birds. The bird will lose condition, show signs of discomfort, and have irritated skin.

Use a pyrethrum-based louse powder every four days for two weeks in autumn and winter (Davies, 2016). The eggs won't be killed, so you'll need to reapply it to catch the larvae. Diatomaceous earth powder can be used on birds and in the coop to kill lice. Avoid stress and overcrowding, and make sure the birds can dust bath.

Mites

The northern fowl mite is the one most commonly found in North America. They cause depression and anemia and may even kill chickens if the infestation goes untreated. They look like dirty patches on the vent feathers and elsewhere on the chicken's body. Cockerels are more likely to be affected.

Apply diatomaceous earth to the whole flock every day for seven days and use it continuously in the coop, too (Davies, 2016).

Internal Parasites

Nematodes or cestodes might infect chickens, leading to weight loss, diarrhea, and reduced egg production. Internal parasites include roundworms, hairworms, gape worms, and tapeworms.

Effective treatments are available for these parasites and are usually added to the feed. Test your birds for worms routinely every three months using a home test kit (Davies, 2016). Manage the ground surrounding the coop to prevent worm eggs from building up. Feeders should preferably be hung up, and waterers kept clean.

Viral Diseases

These are difficult to diagnose and treat and can be very serious if left unresolved. Viral diseases can be highly contagious and could affect your entire flock, so it's important to detect them early. Symptoms include skin sores, coughing and sneezing, reduced egg production, nasal and eye discharges, and even paralysis. Vaccines prevent viral diseases, so ensure that your chicks are vaccinated when you purchase them.

Below are some of the most common viral diseases.

Avian Influenza

This is spread by infected wild waterfowl and then infects poultry. This disease is deadly. Symptoms include diarrhea, nasal discharge, edema in the comb and wattles, purple discoloration, coughing and sneezing, swelling, and ruffled feathers.

Fowl Pox

This is extremely contagious and appears as wet pox or dry pox. The bird develops distinctive bumps that resemble warts on the wattle and comb. Egg production decreases, and young birds may have stunted growth.

Newcastle Disease

This is an acute respiratory disease that spreads very quickly through chicken flocks. The virus may also affect the digestive and nervous systems, so there is a wide range of symptoms. It affects both wild and domestic birds, but the latter are more likely to contract it severely.

Bacterial Diseases

Bacterial diseases are caused by a pathogen or host, which multiplies or grows in the body. They are spread by other infected birds, including wild ones, and are frequently contagious. Symptoms may include breathing difficulties or interruptions in egg laying. Swollen faces and sinuses may be caused by colibacillosis or chronic respiratory diseases.

Colibacillosis

This is caused by E. coli and is an acute, fatal septicemia, usually caused by ingesting contaminated feces. The bacterium causes blood poisoning, which can affect the entire body. Symptoms include coughing, sneezing, reduced appetite, increased defecation, stunted growth, and navel infection. Antibiotics are used to treat this disease.

Chronic Respiratory Disease

Chickens usually become infected through direct contact with infected carrier birds or chicken-keeping equipment. It can also be spread from hens to chicks. Younger birds are more severely affected—their growth might be stunted, and they could die. Symptoms include coughing, noisy lungs, sneezing, eye and nasal discharges, and swelling. Antibiotics treat symptoms, but birds remain carriers for life and can spread the disease.

Coryza

This mostly affects semi-adult and adult birds, and it is found in places continuously inhabited by chickens. The disease arrives quickly, and once exposed, affected birds are carriers for life, even if they are asymptomatic. Signs of coryza include facial and sometimes wattle swelling, sneezing, nasal and eye discharges, pink eye, swollen sinuses, and occasional lower respiratory infections. Introducing infected or carrier birds can spread it to your flock. Vaccines are available.

Fowl Cholera

Mature chickens are more susceptible and are infected by symptomless carrier birds. It is spread by wild birds, predators, rodents, pigs, cats, and contaminated equipment or environments. Symptoms include breathing difficulties, fever, reduced appetite, lower egg production, rapid weight loss, and sudden death. Other signs include a swollen face, wattles, and joints. Vaccines are available, but the best line of defense is to ensure sanitary practices. Controlling rodents and keeping the flock away from wild birds are important in preventing fowl cholera.

Necrotic Enteritis

This generally affects young birds 2–12 weeks old (Cruz-Rincon, 2021). Birds become acutely depressed and die within hours. They often have dark, blood-stained diarrhea, reduced appetite, and weight loss. Birds get infected by eating contaminated feces, soil, and other materials. Antibiotic treatment is effective. Flocks kept in crowded conditions are more likely to become infected. Effective deworming protocols will help prevent outbreaks.

Botulism

This is spread through contaminated food, water, carcasses, maggots, decaying organic material, and sometimes beetles, all of which the bird might ingest. Symptoms can range from dullness and sleepiness to leg, wing, and neck paralysis. Birds may die within 12–24 hours (Cruz-Rincon, 2021). The severity of the symptoms depends on how much toxin is ingested. To prevent botulism, remove all animal carcasses immediately, control flies and insects, and keep decaying organic matter away from your chickens. Affected birds should not be given food or water until they are able to lift their heads. Vitamins and antibiotics may help prevent death. Vaccines are available in places prone to botulism.

These diseases are often spread through drinking water, mold, or contaminated surfaces. Consult your vet to get a proper diagnosis and treatment.

Fungal Diseases

Although rare in chickens, fungal diseases are relatively easy to treat. The most common are brooder pneumonia and ringworm. The former usually affects young chicks, which develop respiratory and breathing problems. Ringworm infections are mild and often clear up on their own. If a chicken has a thick, white layer on the comb, it could be hosting ringworms.

SUMMARY

Various health conditions might affect your chickens. Cuts, sores, and minor wounds are easily treated at home. If your chickens develop parasitic infections from mites and lice, then the coop might need to be disinfected and treated as well. It's also necessary to deworm chickens from time to time. Fungal diseases are normally not serious in chickens and can be treated easily.

More serious bacterial and viral infections are often very contagious and could affect your entire flock if left untreated. Vaccination helps prevent some of the more serious diseases, while your vet can prescribe antibiotics for affected birds. Your birds might be carriers once they have been infected with certain bacteria or viruses.

In the next chapter, you'll find out how to make money from your chicken-keeping efforts.

CHAPTER 8

Cash Chickens—Not Cows

My first business deal was with my mother. I invested in chickens. I sold the eggs to my mother.

— JOEL MCCREA

Raising chickens on a small scale and selling their eggs and meat won't make you a millionaire—and you might not even break even. If you are able to scale up your business, you can automate many aspects and save by buying feed in bulk. Selling the production from your chickens can help you cover your costs, making this a cost-effective hobby. It's also a good way of disposing of excess eggs.

Your financial goal for your hobby farm should be to cover your costs. Once you are up and running, feed should be your only cost unless you amass enormous vet bills, which is unlikely.

MAKING MONEY FROM YOUR CHICKENS

You can sell several products from your chickens.

Fresh Eggs

You'll very likely end up with more eggs than you and your family can eat, so it's a good idea to sell the excess. Don't sell at or below supermarket prices—these eggs are organic and nutritious, with deep yellow yolks, and they're worth paying a little extra for.

Register at your local farmer's market so you have an opportunity to network with other stallholders. Put up a sign at the end of your driveway, or put the word out among your neighbors. Remember to tell people how they can obtain your eggs. Use recycled egg boxes to save costs.

Factors to Consider When Selling Eggs

- Comply with local laws. You might need a permit to sell eggs from home. Make sure you are compliant with any local laws before you sign up for a farmer's market stall.
- It can be difficult to find customers, especially if there are already several chicken keepers in your area.
- Crunch the numbers. How many chickens do you have, how many eggs do they lay weekly, and what does it cost to feed and care for them? How many eggs do you need to sell to cover your production costs? If you want to increase your profits, you need to decide how you will do that. Can you get cheaper,

quality feed? Do you need more chickens or another breed? How much should you charge?
- What will you do when your chickens stop laying?
- Dealing with customers can be tricky. You'll always get those who ask odd questions.

Pricing Your Eggs

How much can you charge for eggs in your area? Most people sell their eggs for far less than grocery stores do, but, considering that people usually purchase eggs from backyard chickens because they are tastier and often organic, you could charge the same—or even a bit more. Most people offer a discount if buyers return their egg cartons, which makes sense as they can then reuse them. There are many online forums covering this topic, so do your research.

Calculate your prices on an annual basis to get an average cost. This will also then take into account things like seasonal egg production, older hens that stop laying, and so on. It's probably best to work on an average of 200 eggs per year—although, if you have older hens, your production will probably be closer to 150 per year. Allow for 1/4 a pound of feed per day for adult chickens (Hancock, n.d.).

The cost of feed per dozen eggs can be calculated using the following formula:

Cost = feed cost x 0.25 (lbs per day) x 365 40lb 200 (eggs per year) x 12 (eggs) (Hancock, n.d.) The cost of chicken feed varies between chicken keepers, as some choose to buy more expensive feed than others. To work out how much your feed costs per dozen eggs, multiply the cost of a 40 lb bag x 0.136 (Hancock, n.d.).

For example, a $50 bag of feed would equal $6.80 in feed cost per dozen eggs.

Remember, your pricing probably won't include set-up costs like building the coop and purchasing feeders, waterers, nest boxes, and roosting bars. There are some items that might raise the cost of producing your eggs. These include:

- treats like corn, scratch grains, or mealworms
- raising chicks and feeding them chick starter feed before they start laying
- taxes
- cartons and labels
- broken eggs or eggs the chickens eat
- laying reductions
- health issues—any vet and medical costs, in addition to reduced laying as a result of ill health
- chicken losses
- broody hens—about 10% go broody, and they don't lay for eight weeks
- soft shell or small eggs (Hancock, n.d.)

If your costing includes these, you could find that your production costs escalate by as much as 30% (Hancock, n.d.).

Remember that it's essential to educate your customers to encourage them to buy your eggs regardless of the price tag. Explain why they should buy them over store-bought ones.

Discuss how you raise your chickens and how fresh and nutritious the eggs are.

Fertilized Eggs

If you have a rooster, sell fertilized eggs to other chicken keepers. Many chicken keepers are looking for unique birds, so if you have special breeds, this might be a good source of extra income.

Day-Old Chicks

Selling chicks to chicken keepers who want to increase the size of their flocks can create a good revenue stream. Besides offering day-old chicks in spring, which is the norm, sell them in fall as well, as chicken keepers may want to make up for any losses. If your chicks are a common breed and not sexed, you won't be able to charge much, but if you are able to sex them, you could ask more. Unfortunately, you might need to dispatch unwanted roosters. Raise the chicks in an incubator. If you sell them as soon as they've hatched, you won't need to incur brooder and heating costs either.

Pullets

Pullets that are about a year old will start laying imminently, so you could sell them to chicken keepers too. The advantage is that you can easily determine the birds' genders at this age, and you know they'll start laying fairly soon, which gives you a marketing advantage.

Stewing Hens

Once your hens have stopped laying, sell them for stew. They won't be particularly tender, but buyers will know how they were raised.

Meat Birds

There's a growing demand for pasture-raised chickens, so now is a good time to raise and sell broilers. As mentioned previously, they mature quickly, so you will soon get a return on your investment. Sell them as processed chickens or live birds that buyers can slaughter themselves. If your chickens eat organic diets or are pasture-raised, you can ask for higher prices per pound.

Guinea Fowl

People are fascinated by guinea fowl, and you can sell their eggs, meat, and feathers. They can be raised alongside your chickens too. This is very advantageous as they are great predator alarms and also help control ticks. Invest in good breeding stock for keets (guinea fowl chicks), so you have more color variations.

Ornamental Feathers

Heritage chicken breeds—especially roosters—have feathers in a range of gorgeous colors. Crafters love them, so collect feathers during the molt and sell them online. They can be used to make earrings or fishing lures, which you can also sell.

Chicken Manure

This is a very popular, nitrogen and phosphorous rich soil amendment that is sought-after by organic gardeners and growers. After being aged for a few weeks, it is ready for sale.

Collect and bag what falls beneath your chickens' roosts, and you'll soon have a new income stream.

Homemade Chicken Feed

Commercial feeds are expensive—and the quality isn't always great. If you make your own chicken feed and find it beneficial, bag and market it to sell to other chicken keepers or small-scale farmers.

You can also make chicken treats and toys, as people love giving their pets something special.

Sell Show Birds at Auction

Consider attending an auction if you have several birds you want to move quickly.

Write About Chickens Online

If you want to share your chicken experiences and knowledge, write about your hobby. Receive payment for recommending products on your own blog or writing articles for others.

Selling Chicken Coops

If you make your own chicken coops, then you can take advantage of this skill. Not all chicken keepers want to build their own coops, so you will have a market for your chicken houses. People want cute and unusual coops or ones that match their house, so you could provide custom-made options.

Rent Out Chicks to Education Centers

If you live near a primary school, consider renting out an incubator so the students can watch chicks hatch. This will save the school money, as they won't need to buy their own incubator. You'll also be able to use your unused supplies, thus saving money.

SUMMARY

As you can see, there are several ways your chicken-keeping hobby could raise extra money for your household. Even if you only cover the cost of your feed, it will be worth the effort. Over time, you might develop a big enough market for your chicken-related products to scale up and have a larger operation.

The idea, however, is to have fun raising these quirky, often hilarious creatures, not to mention enjoying the benefits of fresh eggs, nutritious meat, and beautiful feathers.

Afterword

Raising backyard chickens can be fun and comes with numerous benefits, whether you want fresh eggs, tasty meat, unusual pets, or to teach your children about where their food comes from. Backyard chickens also provide pest and weed control services in the garden, a steady source of organic manure, and feathers for craft and sewing projects.

Decide beforehand why you want to keep chickens, as this will affect your decisions. Choose a breed suitable for your goals and adapted to your local conditions, especially if you live in a particularly hot or cold region. Show chickens at poultry exhibitions, although you may need to invest in unique breeds to do so.

Find out whether your local laws allow you to keep chickens and if there are any limitations as to their number, and whether you can keep roosters. If you intend to raise meat chickens, establish whether you are allowed to process them yourself or if you will need to use processing facilities.

AFTERWORD

You can start off your backyard chicken journey by hatching out fertilized eggs, raising day-old chicks, or buying pullets or mature birds. Some options require special equipment like incubators, brooders, and heat sources. If you want laying birds, purchasing older hens will enable you to get eggs faster. Bear in mind the life cycle of chickens and decide ahead of time how you will deal with birds that become too old to lay.

It's essential that your chickens have a secure, well-constructed coop in which to sleep overnight or take shelter during storms or in cold winters. It must be predator-proof and include nest boxes and roosting bars. Doors, windows, and ventilation should be properly secured so that predators cannot access your hens. Good fencing that is at least 3–6 feet high is essential, the height depending on what predators you have and whether your hens are inclined to fly.

The coop should accommodate your entire flock to limit aggressive behavior and prevent health issues. There should be outdoor access, either in the form of a run or opportunities to free-range on your property. Fence off areas you don't wish the chickens to access. They need a place to dust-bathe, together with plenty of clean, fresh water, which should be replenished regularly.

Feeding your chickens a nutritious, balanced feed will maintain their health, ensuring continuous egg production or tasty meat, depending on your chicken goals. Feed requirements change as the birds grow. Broilers (meat chickens) grow extremely fast and can be slaughtered as early as eight weeks of age.

AFTERWORD

Egg-laying chickens require specific nutrients to ensure egg production and superior egg quality. Certain supplements, especially calcium in the form of oyster shells or crushed eggshells, are essential to maintain hard eggshells and ensure the health of laying chickens. Chickens can be fed treats in moderation, including mealworms, table scraps, scratch grains, leafy greens, and fruit.

Chickens are sociable birds and have a strictly hierarchical social structure—the pecking order—where certain birds are dominant. Activities are governed by the chickens' places in the pecking order. Roosters usually dominate, but they protect the flock from predators and raise the alarm, and they find good food sources and mate with the hens to produce fertilized eggs.

Chickens are very vocal birds and have 30 different sounds to communicate with one another. They use displays, a sort of bird body language, as a prelude to mating and to indicate territoriality or dominance.

These birds are generally healthy, but if a serious disease takes hold in your coop, it could affect your whole flock. Certain bacterial and viral diseases might not kill your chickens, but they will become carriers after being exposed. Others are deadly. There are simple methods of combating lice and mites, worms, and fungal conditions, but more serious health issues or diseases may need antibiotics or veterinary treatment.

You can make money from your backyard chickens. Although you won't be a millionaire, this can help you cover costs. Besides the obvious avenues of selling eggs and meat, you can also collect and bag manure, sell feathers for craft objects, build and sell chicken coops, make and sell your own chicken feed, or rent an incubator to a primary school so children can see how chicks hatch.

Keeping backyard chickens is a very rewarding experience. You will develop a strong bond with your "girls," and you will discover many things about them. Not only that, you will have a steady supply of eggs, meat, and feathers and will be connected to your food in a very unique and personal way. You will have garden helpers that control weeds and insects and provide you with loads of free manure. Your children will have unusual pets that ride on their shoulders, take an interest in whatever they are doing, and teach them about where their food comes from.

* * *

LEAVE A REVIEW

If you have enjoyed this book, remember to leave a review.

Bibliography

Al. (2022, May 17). *The chicken cycle: The benefits of gardening with chooks.* Alikats. https://alikats.eu/blog/sustainability/the-chicken-cycle-the-benefits-of-gardening-with-chooks/

Amy. (2018, February 18). *Should you sell chicken eggs? A few things to consider.* A Farmish Kind of Life. https://afarmishkindoflife.com/sell-chicken-eggs/

Andrews, C. (n.d.). *Choosing hatching eggs: four steps to successful incubation.* Raising Happy Chickens. https://www.raising-happy-chickens.com/choosing-hatching-eggs.html

Arcuri, L. (2009). *Easy chicken care for your small farm.* The Spruce. https://www.thespruce.com/daily-and-monthly-chicken-care-tasks-3016823

Arcuri, L. (2023, January 20). *Tips for collecting and cleaning chicken eggs.* The Spruce. https://www.thespruce.com/collect-clean-and-store-chicken-eggs-3016828

Arouri, L. (2012). *What should I feed my chickens?* The Spruce. https://www.thespruce.com/feeding-your-chickens-or-laying-hens-3016556

Backyard chicken health issues: Symptoms, treatment, and prevention. (n.d.). Caring Pets. https://www.caringpets.org/how-to-take-care-of-a-backyard-chicken-hen/health-issues/#bacterial

Balam, G. (n.d.). *How to raise chickens: Answers to popular questions.* Purina Mills. https://www.purinamills.com/chicken-feed/education/detail/how-to-raise-chickens-answers-to-popular-questions#:~:text=On%20average%2C%20a%20laying%20hen

Barth, B. (2016, March 16). *The secret of chicken flocks' pecking order.* Modern Farmer. https://modernfarmer.com/2016/03/pecking-order/

Beginner's guide to raising backyard chickens. (2018, May 25). The Happy Chicken Coop. https://www.thehappychickencoop.com/raising-chickens/

Bentoli. (2017, August 15). *6 common chicken problems you can fight with*

BIBLIOGRAPHY

proper nutrients. Bentoli. https://www.bentoli.com/chicken-problems-common/

Biggs, P. (n.d.a). *Can I raise backyard chickens in my area?* Purina Mills. https://www.purinamills.com/chicken-feed/education/detail/can-i-raise-backyard-chickens-in-my-area

Biggs, P. (n.d.b). *What to feed chickens: Chicken treats to feed and avoid*. Purina Mills. https://www.purinamills.com/chicken-feed/education/detail/what-to-feed-chickens-chicken-treats-to-feed-and-avoid

Biggs, P. (2018). *How long do chickens lay eggs?* Purina Mills. https://www.purinamills.com/chicken-feed/education/detail/how-long-do-chickens-lay-eggs-goals-for-laying-hens

Brahlek, A. (2022, September 28). *How to prep your chickens for a hurricane and other severe weather event*. Grubbly Farms. https://grubblyfarms.com/blogs/the-flyer/how-to-prep-your-chickens-for-a-hurricane

Brainy Quotes. (n.d.). *Chicken quotes*. BrainyQuote. https://www.brainyquote.com/topics/chicken-quotes

Caley, N. (2021, December 1). *Count your chickens*. Pet Business. https://www.petbusiness.com/pets-and-products/count-your-chickens/article_d7b8965c-4b20-11ec-ab27-4b7bd06a1292.html

Caring for baby chicks: What to do once they arrive. (n.d.). Purina Mills. https://www.purinamills.com/chicken-feed/education/detail/caring-for-baby-chicks-what-to-do-once-they-arrive

Chicken bullying: How to stop them. (2021, May 12). *pecking each other*. The Happy Chicken Coop. https://www.thehappychickencoop.com/chicken-bullying/

Chicken life cycle (learn the 4 key stages). (2018, May 3). The Happy Chicken Coop. https://www.thehappychickencoop.com/chicken-life-cycle/

Common Chicken Eye Problems. (n.d.). Agriculture Site. https://agricsite.com/chicken-eye-problems/

Cooper, T. (2020, September 14). *7 chicken coop basics that your chickens need*. Backyard Poultry. https://backyardpoultry.iamcountryside.com/coops/chicken-coop-basics-chickens-need/

Cosgrove, N. (2021, June 24). *How much does it cost to raise chickens? (2023 Price Guide)*. Pet Keen. https://petkeen.com/cost-to-raise-chickens/

Crank, R. (2019, July 18). *How to raise free-range chickens*. Backyard Poultry. https://backyardpoultry.iamcountryside.com/chickens-101/how-to-raise-free-range-chickens/

BIBLIOGRAPHY

Crank, R. (2022, April 19). *The pros and cons of letting chickens roam free.* Pet Helpful. https://pethelpful.com/farm-pets/Pros-and-Cons-of-Letting-Chickens-Free-Roam

Cruz-Rincon, S. (2021, April 2). *Common bacterial diseases in backyard chickens.* Veterinary Partner. https://veterinarypartner.vin.com/default.aspx?pid=19239&id=10048768

Davies, G. (2016, December 2). *Common poultry parasites of backyard hens.* The Veterinary Nurse. https://www.theveterinarynurse.com/review/article/common-poultry-parasites-of-backyard-hens

Doug. (2017, February 9). *9 rules for hatching eggs.* The Chicken Tractor. https://www.thechickentractor.com.au/9-rules-for-hatching-eggs/

Egg Binding. (n.d.). British Hen Welfare Trust.https://www.bhwt.org.uk/hen-health/health-problems/egg-binding/#:~:text=A%20warm%20bath%20followed%20by

8 Ways to Make Money With Your Backyard Chickens. (n.d.). Strombergs. https://www.strombergschickens.com/blog/8-ways-to-make-money-with-your-backyard-chickens/

Essie, K. (2019, October 9). *20 Real-life funny chicken stories.* Hobby Farms. https://www.hobbyfarms.com/20-real-life-funny-chicken-stories/

Flocks guardian: Guard animals to protect chickens. (2022, March 7). The Happy Chicken Coop. https://www.thehappychickencoop.com/flocks-guardian-guard-animals-to-protect-chickens/

Finn, A. (2021, July 22). *120 Chicken quotes to make you appreciate them.* Quote Ambition. https://www.quoteambition.com/chicken-quotes/

Garman, J. (2021, May 5). *Treating livestock and chicken eye problems.* Backyard Poultry. https://backyardpoultry.iamcountryside.com/feed-health/treating-livestock-and-chicken-eye-problems/

Guest Author. (2020, March 24). *How much space do chickens need?* K&H Pet Products. https://khpet.com/blogs/farm/how-much-space-do-chickens-need

Hamre, M., & Phillips, H. (2021). *Raising chickens for meat.* University of Minnesota. https://extension.umn.edu/small-scale-poultry/raising-chickens-meat#outdoor-density-for-free-range-broilers-1580715

Hancock, A. (n.d.). *How much are your eggs worth?* Happy Wife Acres. https://www.happywifeacres.com/how-much-are-your-eggs-worth/

Harm, S. (2019, April 10). *Bawk to the future: How backyard chicken keeping began as a war effort.* austintexas.gov. https://www.austintexas.-

BIBLIOGRAPHY

gov/blog/bawk-future-how-backyard-chicken-keeping-began-war-effort

Herd, R. (2016, July 27). *Your easy-to-follow guide to chicken chores.* Hobby Farms. https://www.hobbyfarms.com/your-easy-to-follow-guide-to-chicken-chores/

Hotaling, A. (2016, June 13). *Raising show chickens.* Hobby Farms. https://www.hobbyfarms.com/raising-show-chickens/

How cold is too cold for my chickens? (2021, March 17). The Happy Chicken Coop. https://www.thehappychickencoop.com/how-cold-is-too-cold-for-my-chickens/#:~:text=Chickens%20can%20survive%20quite%20well

How many people have gotten chickens in the past five years? (2022, March 30). The Happy Chicken Coop. https://www.thehappychickencoop.com/how-many-people-have-gotten-chickens-in-the-past-five-years/

How much room do chickens need? (2021). The Happy Chicken Coop. https://www.thehappychickencoop.com/how-much-room-do-chickens-need/

Hudson, J. (2022, November 23). *How cold is too cold for your chickens?* Chicken Scratch. https://cs-tf.com/how-cold-is-too-cold-for-chickens/#:~:text=Cold%20weather%20chickens%20can%20withstand,to%20about%20ten%20degrees%20Fahrenheit).

Hynson, J. (2017, December 11). *7 benefits of gardening with chickens.* The Thrifty Homesteader. https://thriftyhomesteader.com/gardening-with-chickens/

Jacob, J. (n.d.a). *Normal behaviors of chickens in small and backyard poultry flocks* poultry.extension.org. https://poultry.extension.org/articles/poultry-behavior/normal-behaviors-of-chickens-in-small-and-backyard-poultry-flocks/

Jacob, J. (n.d.b). *Raising meat chickens in small or backyard flocks.* poultry.extension.org. https://poultry.extension.org/articles/poultry-management/raising-meat-chickens-in-small-or-backyard-flocks/https://poultry.extension.org/articles/poultry-management/raising-meat-chickens-in-small-or-backyard-flocks/

James. (2022, January 10). *Do chickens need vitamin and mineral supplements?* Learn Poultry. https://learnpoultry.com/chickens-need-vitamin-mineral-supplements/

Josephson, A. (2018, May 18). *The economics of raising chickens.* SmartAsset.

https://smartasset.com/personal-finance/the-economics-of-raising-chickens

Lesley, C. (2020, July 2). *The complete life cycle of a chicken explained*. Chickens and More. https://www.chickensandmore.com/life-cycle-of-a-chicken/

Lesley, C. (2022a, May 5). *Common chicken health problems*. Almanac. https://www.almanac.com/common-chicken-health-problems

Lesley, C. (2022b, December 29). *Raising chickens 101: How to build a chicken coop*. Almanac. https://www.almanac.com/raising-chickens-101-how-build-chicken-coop

MacLean, K. (n.d.). Understanding *Local laws for raising backyard chickens*. Pete and Gerry's Organic Eggs. https://www.peteandgerrys.com/blog/local-laws-for-raising-backyard-chickens

Manion, B. J. (2016, March 27). *Daily, weekly, monthly, and semi-annual chores for caring for your chickens*. Dummies. https://www.dummies.com/article/home-auto-hobbies/hobby-farming/chickens/daily-weekly-monthly-and-semi-annual-chores-for-caring-for-your-chickens-204282/

McCrae, B., & Baker, B. (2022, November 2). *Common backyard Chicken behaviors*. Alabama Cooperative Extension System. https://www.aces.edu/blog/topics/farming/common-backyard-chicken-behaviors/

Meredith. (2019, April 1). 24 *Features on a predator-proof chicken coop*. Backyard Chicken Project. https://backyardchickenproject.com/predator-proof-chicken-coop/

Show Day Prep for Chickens. (2020, June 22). Meyer Hatchery Blog. https://blog.meyerhatchery.com/2020/06/show-day-prep-for-chickens/

Mike (2021, April 5). *How high should a chicken fence be to keep chickens in and predators out?* Outdoor Happens. https://www.outdoorhappens.com/how-high-should-a-chicken-fence-be-to-keep-chickens-in-and-predators-out/

Choosing chicken breeds. (2015, February 18) Miller Manufacturing Company Blog. https://www.miller-mfg.com/blog/choosing-chicken-breeds/

Manitoba Agriculture. (n.d.). *Cracked eggs and your small flock of laying hens.* https://www.gov.mb.ca/agriculture/livestock/production/poultry/cracked-eggs-and-your-small-flock-of-laying-hens.html

Moore, J. (2018, June 8). *Flock dynamics: a guide to the social hierarchy of chickens.* The Country Smallholder. https://thecountrysmallholder.-

BIBLIOGRAPHY

com/poultry/flock-dynamics-a-guide-to-the-social-hierarchy-of-chickens-8257610/

Mormino, K. S. (n.d.). *Backyard chicken keeping information and advice.* The Chicken Chick. https://the-chicken-chick.com/

Mormino, K. S. (2013a, April 2). *When to move chicks from brooder to chicken coop.* The Chicken Chick. https://the-chicken-chick.com/when-to-move-chicks-from-brooder-to/

Mormino, K. S. (2013b, July 4). *11+ tips for predator-proofing chickens.* The Chicken Chick. https://the-chicken-chick.com/11-tips-for-predator-proofing-chickens/

Mormino, K. S. (2016, March 31). *How much heat do chicks really need? Think like a mother hen!* The Chicken Chick. https://the-chicken-chick.com/how-much-heat-do-chicks-really-need/

Morning Chores Staff. (2018, December 12). *How to go about choosing the perfect chicken breeds for you.* Morning Chores. https://morningchores.com/choosing-a-chicken-breed/

My Favorite Chicken. (2020, July 10). *A beginner's guide: How to build a chicken coop.* https://myfavoritechicken.com/2020/07/10/a-beginners-guide-how-to-build-a-chicken-coop/

Noyes, L. (2019, October 1). *14 ways to make money from your backyard chickens.* Rural Sprout. https://www.ruralsprout.com/make-money-from-backyard-chickens/

Ohio State University. (2017, August 17). *Chicken breed selection.* https://ohioline.osu.edu/factsheet/anr-60

Rachael. (2020, October 24). *What age chicken to buy - A beginner's guide to getting the right chickens for you.* Dine a Chook. https://www.dineachook.com.au/blog/what-age-chicken-to-buy-a-beginners-guide-to-getting-the-right-chickens-for-you/

Rasmus. (2021, July 6). *How to discipline your chickens.* Poultry Parade. https://poultryparade.com/how-to-discipline-your-chickens/

Reddy, D. (2021, February 4). *100 best chicken quotes, sayings and phrases.* Kidadl. https://kidadl.com/quotes/best-chicken-quotes-sayings-and-phrases

Roberts, J. (2019, July 18). *Feeding oyster shell to backyard chickens (guide).* Know Your Chickens. https://www.knowyourchickens.com/oyster-shells-for-chickens/

Roeder, M. (2019). *21-day guide to hatching eggs.* Purina Mills.

https://www.purinamills.com/chicken-feed/education/detail/hatching-eggs-at-home-a-21-day-guide-for-baby-chicks

Rosen, R. (2020, September 15). *Chicken myths you're still believing (and why you shouldn't!)*. Chicken Check In. https://www.chickencheck.in/blog/chicken-myths-you-still-believe/

Sawyers, D. (2020, January 2). *Sexing chicks: How to determine gender of your chickens*. Freedom Ranger Hatchery. https://www.freedomrangerhatchery.com/blog/sexing-chicks-how-to-determine-gender-of-your-chickens/

Schneider, A. (2018, July 26). *Six benefits of owning chickens*. Acreage Life. https://www.acreagelife.com/hobby-farming/six-benefits-of-owning-chickens

7 benefits to raising backyard chickens. (n.d.). Stromberg's. https://www.strombergschickens.com/blog/7-benefits-to-raising-backyard-chickens/

Sextro, M. (2019, October 13). *14 chicken supplements and how they can help!* Backyard Chicken Project. https://backyardchickenproject.com/14-chicken-supplements-and-how-they-can-help/

Should I have a rooster? Learn the pros and cons. (2019, May 15). The Happy Chicken Coop. https://www.thehappychickencoop.com/should-i-have-a-rooster/

Smith, K. (2020a, July 16). *How to stop your flock's bad habits*. Backyard Chicken Coops. https://www.backyardchickencoops.com.au/blogs/learning-centre/how-to-stop-your-flocks-bad-habits

Smith, K. (2020b, July 22). *What age chicken should I buy?* Backyard Chicken Coops. https://www.backyardchickencoops.com.au/blogs/learning-centre/what-age-chicken-should-i-buy

Space allowances in housing for small and backyard poultry flocks. (n.d.). Poultry Extension. https://poultry.extension.org/articles/getting-started-with-small-and-backyard-poultry/housing-for-small-and-backyard-poultry-flocks/space-allowances-in-housing-for-small-and-backyard-poultry-flocks/

Steele, L. (2015). *Spice up your chicken keeping for better flock health*. Fresh Eggs Daily https://www.fresheggsdaily.blog/2015/05/spice-up-your-chicken-keeping-for.html

Steele, L. (2020, February 23). *Oregano for chickens: Build stronger immune systems*. Backyard Poultry. https://backyardpoultry.iamcountryside.com/feed-health/oregano-for-chickens-build-stronger-immune-systems/

BIBLIOGRAPHY

Steele, L. (2023, February 10). *Raising chickens 101: Choosing the right chicken breeds.* Almanac. https://www.almanac.com/raising-chickens-101-choosing-chicken-breeds

Telkamp, M. (n.d.). *Why chickens roost.* HGTV. https://www.hgtv.com/outdoors/gardens/animals-and-wildlife/why-chickens-roost

Telkamp, M. (2019). *What do chickens eat?* HGTV. https://www.hgtv.com/outdoors/gardens/animals-and-wildlife/what-do-chickens-eat

The Editors. (2021, September 22). *Raising chickens 101: When chickens stop laying eggs.* Almanac. https://www.almanac.com/raising-chickens-101-when-chickens-stop-laying-eggs

Thornberry, F. D. (n.d.). *How to produce broilers and roasters for show.* http://posc.tamu.edu/wp-content/uploads/sites/20/2012/08/L5431.pdf

What Is My Chicken Doing? Chicken Behavior 101. (2021, July 13). The Pioneer Chicks. https://www.thepioneerchicks.com/what-is-my-chicken-doing-chicken-behavior-101/

Winger, J. (2013, April 3). *What not to feed chickens.* The Prairie Homestead. https://www.theprairiehomestead.com/2013/04/7-things-you-shouldnt-feed-your-chickens.html

Winger, J. (2020, March 12). *Beginner's guide to raising laying hens.* The Prairie Homestead. https://www.theprairiehomestead.com/2020/03/raising-laying-hens.html

Yin, S. (2016, September 8). *Why raise backyard poultry? 8 indisputable reasons why you should.* NOBOWA. https://nobowa.com/benefits-raising-backyard-chickens/

809 Creative Chicken Names

FROM TRADITIONAL, ECCENTRIC, AND FUNNY ALL THE WAY TO MODERATELY INAPPROPRIATE: UP YOUR CHICKEN-NAMING GAME WITH THIS HELPFUL COMPILATION

Introduction

Welcome to the wonderful world of chicken names! Naming your feathered friends is a fun and exciting way to express your creativity and add personality to your flock. Whether you have a few chickens in your backyard or a whole coop full, finding the perfect names can be a delightful challenge and is one of the most fun chicken activities of all poultry-related tasks. In this book, you'll discover 809 suggestions and ideas to inspire your chicken-naming adventure. Names are categorized into Chapters for more accessible selection. We've got traditional, all kinds of funny, classic, eccentric, goofy, "punny," and some bordering on offensive – there is something here for every Crazy Chicken Lady, and every chicken!

Every Crazy Chicken Lady, and every chicken has a unique personality and distinct characteristics. Sometimes people name their chickens based on physical traits like color, feather pattern, or size. Some earn their names from their personality, disposition, or a particular quirk. One Magnificent Crazy Chicken Lady I know has a particularly "rough" looking

INTRODUCTION

rooster she calls Keith Richards. Perfection. Another has named a good portion of her flock after the cast of one of her favorite TV shows. I've got one that likes to chase cows for some reason. Her name is Cow Dog.

But not everyone goes for the funny or sarcastic angle – you could just as easily choose classic leading ladies; Greta Garbo, Elizabeth Taylor or Sofia Loren. Or stick with puns – Feather Locklear, Chick Norris, Tu-Peck. The list goes on – so let's get cracking and find the perfect name for your cluckers!

CHAPTER 1
Traditional

Do not count your chickens before they are hatched.

— AESOP

Why did the chick cross the road?
To get to the other side.

- Mrs./Miss/Mr. Peckers
- Chicken Little
- Tweety
- Foghorn Leghorn
- Woodstock
- Fluffy
- Puffy
- Waddles
- Diva
- Pecky

- Heihei
- Lady Kluck
- Ginger
- Torchic
- Alan-a-dale
- Miss Prissy
- Sheldon
- Panchito Pistoles
- Babs
- Chickaletta
- Ernie the Giant Chicken
- Roy Rooster
- Fowler
- Chanticleer
- Robot Chicken
- Buck Cluck
- Clara Cluck
- Runt of the Litter
- Goldie
- Blondie
- Sunny
- Raven
- Cotton
- Snow
- Chocolate
- Vanilla
- Strawberry
- Butterscotch
- Caramel
- Cinnamon
- Honey

- Salty
- Sweet
- Sour
- Bitter
- Shadow
- Midnight
- Chickie Chickie
- Cluck Cluck
- Eggbert
- Gertrude McFuzz
- Peckov
- Cluckov
- Chickov
- Roosterofsky
- Chickovsky
- Angry Bird

CHAPTER 2
A Little Bit Country...

CHICKEN SONGS:

The Chicken In Black – Johnny Cash
Chicken Fried – Zac Brown Band
Dixie Chicken – Little Feat
Sic 'Em On A Chicken – Zac Brown Band
Chicken Train – The Ozark Mountain Daredevils

- Boots
- Lasso
- Barb Wire
- Cowboy Coffee
- Horseshoe
- Hoof Pick / Hoof "Peck"
- Spurs
- Tom Sawyer
- Huck Finn
- Davy Crocket

- Daniel Boone
- James Bowie
- Wild Bill Hickock
- Calamity Jane
- Annie Oakley
- Belle Starr
- Pearl Hart
- Buffalo Bill
- Billy the Kid
- Wyatt Earp
- Doc Holiday
- Sam Bass
- Dale Earnhardt
- Dale Jarrett
- Daryll Waltrip
- Michael Waltip
- Dick Trickle
- Clint Eastwood / "Cluck" Eastwood
- Bonanza
- John Wayne / The Duke
- Rooster Cogburn
- Big Jake
- Tonto
- The Long Ranger
- Miss Kitty
- Chester
- Zorro
- Aunt Bea
- Barney Fife
- Opie
- Katie Elder

- Laura Engels
- Mr. Ed
- Howdie Doodie
- Merle Haggard
- Waylon Jennings
- Buck Owens
- Hank Williams
- Johnny Cash
- Georg Strait
- Willie Nelson
- Porter Wagner
- Conway Twitty
- Garth Brooks
- Rodney "Cockington" Carrington
- Urban Cowboy
- Rhinestone Cowboy
- Dolly Parton
- Tammy Wynette
- Loretta Lynn
- Reba McEntire
- Shania Twain
- Faith Hill
- Kelly Clarkson / Kelly "Cluck"-son
- Tanya Tucker / Tanya "Clucker"
- Teri Clark / Teri "Cluck"
- Crystal Gale
- Patsy Cline
- Pam Tillis
- Lorrie Morgan
- The Judds (Wynona, Ashley, Naomi)
- Elvira

- Huckleberry
- Mustang Sally
- Casey Tibbs
- Larry Mahan
- Jim Shoulders
- Ty Murray
- Roy Duval
- Does your chicken chase cows? Cattle Guard, Wrangler, or Cow Dog.

CHAPTER 3
A Little Bit Rock 'n' Roll...

Life was just a tire swing... Blackberry pickin', eatin' fried chicken.

— JIMMY BUFFET

CHICKEN SONGS:

Ain't Nobody Here but Us Chickens – Louis Jordan
Chicken Farm – Dead Kennedys
5-Piece Chicken Dinner – Beastie Boys

- Elvis
- Jerry Lee Lewis
- Freddy Mercury
- Cher
- Madonna

- Tina Turner
- Joni Mitchell
- ACDC
- Quiet Riot
- Fleetwood Mac
- Creedence Clearwater Revival
- Joan Jett
- Cyndi Lauper
- Slash
- Sting
- David Bowie / David "Crowie"
- Mick Jagger / "Chick Jagger"
- Ozzy Osbourne
- Melissa Etheridge
- Iggy Pop
- Stevie Nicks
- Janis Joplin
- Kurt Cobain
- Courtney Love
- Chuck Berry
- Pat Benatar
- Keith Richards
- Alice Cooper
- James Hetfield
- Axl Rose
- Janice Joplin
- Steven Tyler
- Rock-a-bye-Billie
- Olivia Newton John

CHAPTER 4
Thug Life / Hip Hop / R&B

Mmm...Fried chicken, fly vixen /

Give me heart disease but need you in my kitchen

— NAS

CHICKEN SONGS:

> Do the Funky Chicken – Rufus Thomas
> I Move Chickens – Gucci Mane

- TuPac / "Tu-Peck"
- Notorious B.I.G. / Biggie Smalls
- P Diddy / Puff Daddy / Puffy "Combs"
- Snoop Dogg
- Busta Rhymes
- Dr. Dre

- Ice Cube
- Wu-Tang Clan
- Jay-Z
- 50 Cent
- Sir Mix-A-Lot / Sir Chicks-A-Lot / Sir Clucks-A-Lot / Sir Crows-A-Lot
- Vanilla Ice
- MC Hammer
- Kanye
- Beyonce (Destiny's Child)
- Kelly Rowland (Destiny's Child)
- Michelle Williams (Destiny's Child)
- Lisa "Left Eye" Lopez (TLC)
- Tionne "T-Boz" Watkins (TLC)
- Rozonda "Chilli" Thomas (TLC)
- Mary J Blige
- Missy Elliott
- Lauryn Hill
- Queen Latifah
- Foxy Brown
- Lil' Kim
- Cardi B
- Nicki Minaj
- Lizzo

CHAPTER 5
More Female Vocal Powerhouses

Eat fried chicken every day, as the angels go sailing by.

— ELLA FITZGERALD

- Aretha Franklin
- Etta James
- Ella Fitzgerald
- Billie Holiday
- Nina Simone
- Carole King
- Donna Summer
- Diana Ross
- Karen Carpenter
- Barbara Streisand
- Annie Lennox
- Gladys Knight
- Amy Winehouse

- Celine Dion
- Mariah Carey
- Whitney Houston
- Christina Aguilera
- Brittany Spears
- Lady Gaga
- Adele

CHAPTER 6
Overly Aggressive Roosters

I have a painting where somebody's holding a chicken, and underneath the chicken is somebody's head.

— JEAN-MICHEL BASQUIAT

Why don't chickens like people?
They beat eggs.

- Trouble
- The Plague / Bubonic
- Lucifer / "Roocifer"
- Scarface
- Vito Corleone
- Michael Vick
- Harvey Weinstein
- Bill Cosby
- Jeffrey Epstein

- Prince Geoffrey
- OJ Simpson
- Charlie Manson
- John Wayne Gacy
- Ted Bundy
- Jefferey Dahmer
- David Berkowitz
- Hannibal Lecter
- Bianchi and Buono
- Richard Ramirez
- HH Holmes
- Jack the Ripper
- Buffalo Bill
- Pablo Escobar
- Hitler
- Osama Bin Laden
- Ivan the Terrible
- Vlad the Impaler
- Benito Mussolini
- Joseph Stalin
- Vladimir Putin
- Muammar Gaddafi
- Kim Jong-un

CHAPTER 7
Violent Hens

"When your mama was the geek, my dreamlets", Papa would say, "she made the nipping off of noggins such a crystal mystery that the hens themselves yearned toward her, waltzing around her, hypnotized with longing."

— KATHERINE DUNN

- Susan Atkins
- Attila the Hen
- Morgan le Fay
- Miranda Priestly
- Wednesday Addams
- Veruca Salt
- Mommy Dearest
- Mama Fratelli
- Norma Bates

- Annabelle
- Madam Mim
- Carrie
- Calypso
- Delores Claiborne
- Annie Wilkes
- Ghislaine Maxwell
- Lizzie Borden
- Bloody Mary
- Aileen Wuornos
- Red Sparrow
- Red Queen
- Daenerys
- Cersei Lannister
- Tanya Harding
- Hela
- Firestarter
- Maleficent

CHAPTER 8
For the Fighters

What did the counsellor say to the egg?
Say no to crack.

- Muhammad Ali
- Joe Frazer
- Cassius Clay
- Rocky "Cocky" Balboa
- Apollo Creed
- Ivan Drago
- John Wick
- Bruce Lee
- Undertaker
- Birdzilla
- Beth Dutton
- Bellatrix Lestrange
- Harley Quinn
- Ellen Ripley
- Furiosa
- Daenerys Targaryen

- Khaleesi
- Arya Stark
- Cruella de Vil
- The Scarlet Witch
- Nebula
- Poison Ivy
- Queen
- Wicked Witch of the West
- Alien Queen
- Nurse Ratched
- Medusa
- Lady Macbeth
- Lady Tremaine

CHAPTER 9
So Wrong...But So Right...

I was eating in a Chinese restaurant downtown. There was a dish called Mother and Child Reunion. It's chicken and eggs. And I said, I gotta use that one.

— PAUL SIMON

- Breakfast
- Yoko Ono / "Yolko" Ono
- John Denver Omelet
- Consuela Spanish Omelet
- Over Easy
- Poached
- Braised
- Eggo
- Bacon and Eggs
- Hard Boiled
- Soft Boiled

- Deviled
- Salmonella
- Chick-Fil-A
- Boneless
- Piccata
- Sesame
- Hollandaise
- Frittata
- Parmesan
- Noodle
- Barbie Q
- Original Recipe
- Extra Crispy
- Drum Stick
- Hot Wing
- Rotisserie
- Peking Duck
- Dark Meat
- White Meat
- Chicken Nugget
- Stuffin'
- Brunch
- Quiche
- Soup
- Pot Pie
- Fried Steak
- Enchilada

CHAPTER 10
Just Because I Think These are Funny

A Jewish woman had two chickens. One got sick, so the woman made chicken soup out of the other one to help the sick one get well.

— HENNY YOUNGMAN

- Pox
- Ugly Duckling
- Hammock
- Tighty Whitey
- Thong
- Brisket
- Piglet
- Kitten
- Tadpole
- Catfish
- Cheeseburger

809 CREATIVE CHICKEN NAMES

- Muffin
- Pickle
- Lunchmeat
- Taco
- Turducken
- Kung Pao
- Bugs Bunny
- Yosemite Sam
- Hot Tamale
- Flip Flop
- Slippers
- High Heel
- Stiletto
- Rubber Boots
- Birkenstock
- Bullwinkle
- Elmo
- Cat Lady
- Lester Holt
- Tom Brokaw
- Walter Cronkite
- Diane Sawyer
- Katie Couric
- Barbara Walters
- William Dafoe
- Jeff Goldblum
- Jim Carry
- Rick Moranis
- Got a fast chicken? How about Secretariat, or Sea Biscuit? Speedy Gonzales? Too Obvious. Run DMC - better.

- Donkey Kong
- Hulk Hogan
- Rowdy Roddy Piper
- Maybe I Should Get Bangs
- I Should Start a Pod Cast
- Business in Front, Party in the Back
- Superhero Landing
- Negasonic Teenage Warhead
- The Real Housewives of [enter your house/town/farm here]
- The Pope
- The Girl with the Chicken [or Rooster] Tattoo
- Cool Runnings
- I'm Still Going To Use Someone Else's Netflix
- Or just...Netflix
- Hulu
- Peacock

CHAPTER 11
More For Roosters

Why did the rooster never come home to his hen at night?
He was free range.

- Alarm Clock
- Noise Machine
- Noise Maker
- Big Daddy
- Mr. Big
- Colonel Sanders
- Russel "Crowe"
- Tank
- Chuck Norris / "Cluck" Norris / "Chick" Norris
- Mr. T
- Johnny Bravo
- Rico Suave
- Weird Al
- Boy Georg
- Crackhead Carl
- Frankie Doodle Doo

- Lenny "Crow"-vtiz
- "Cluck" Kent
- Kung Fu
- Pecker
- James Bond
- General Tso
- Punky Rooster
- Franklin Delano "Roostervelt"
- Mr. Rogers
- The Big Lebowski / The Dude
- Big Bootie Rudy
- Cock-a-doodle-DON'T
- Barry Mani-"crow"
- "Roo" Hefner
- "Roo" Paul
- Luke "Combs"
- Shawshank
- Stephen King
- Quentin Tarantino
- David Hasselhoff

* * *

If you are enjoying this book, please consider leaving a review.

* * *

CHAPTER 12
Adult Theme

What happens when hens and roosters get together?
It's eggciting.

Why did the chicken cross the road, roll in the mud, and cross again?
Because it was a dirty double crosser.

- Hanky Panky
- Poker in the Front
- Dirty Duck
- Peeping Tom
- John Holmes
- Ron Jeremy / The Hedgehog
- Captain Cock
- Mrs. Pull It
- Mother Clucker
- Cluck It
- What the Cluck
- Cluck Off

- BBC – Big Black Chicken
- Jezebel
- Pink Lady
- Half Baked
- Dazed and Confused
- Pineapple Express
- Drunk and Disorderly
- Drunk in Public

Ever had a Rooster that turned out to be a Hen? How about:

- Bruce "Henner"
- A Rooster Named Sue

CHAPTER 13
Duos, Trios, and Ensembles

We can see a thousand miracles around us every day. What is more supernatural than an egg yolk turning into a chicken?

— S. PARKES CADMAN

- Thelma and Louise
- Elsa and Ana
- Maverick and Goose
- Ann and Nancy Wilson
- Smokey and The Bandit
- Smith and Wesson
- Hannibal Lector and Clarice Starling
- Ike and Tina Turner
- Brad Pitt and Angelina Jolie
- Mickey and Minnie
- Mario and Luigi

- Bonnie and Clyde
- Salt-n-Pepa
- Bill and Ted
- Bert and Ernie
- Beavis and Butthead
- Wayne and Garth
- Jack, Janet, and Chrissy
- Harry, Ron and Hermione
- Buffy, Willow, and Xander
- Captain Kirk, Spock, and McCoy
- Larry, Curly, and Moe
- Chevy Chase, Steve Martin, and Martin Short
- Farrah Fawcett, Jaclyn Smith, and Kate Jackson
- Blossom, Bubbles, and Buttercup
- Judy, Violet and Doralee
- Ferris, Cameron, and Sloane
- Alvin, Simon, Theodore
- Shrek, Fiona, and Donkey
- Snap, Crackle, and Pop
- Frank Sinatra, Dean Martin, and Sammy Davis Jr.
- Carrie, Samantha, Charlotte, and Miranda
- Dorothy, Blanche, Rose and Sofia
- John Lennon, Paul McCartney, Ringo Starr, George Harrison
- Stan, Kyle, Cartman, and Kenny
- Tootie, Blair, Natalie, Jo, and Mrs. Garrett
- Jerry, Kramer, Elaine, Costanza, Newman
- Scary Spice, Sporty Spice, Baby Spice, Ginger Spice, Posh Spice
- Dopey, Doc, Bashful, Sneezy, Happy, Grumpy and Sleepy

- Hank, Peggy and Bobbi Hill, Luanne Platter, John Redcorn, Boomhower, Bill, and Dale
- Charlie Brown, Linus, Lucy, Peppermint Patty, Sally, Schroeder, Franklin, Marcie, Pig-Pen, Woodstock, Snoopy
- Sam Malone, Woody Boyd, Diane Chambers, Rebecca Howe, Carla Tortelli, Norm Peterson, Cliff Clavin, Frasier Crane, Lilith Sternin

CHAPTER 14
Inspired by our Favorite Adult Beverages

Did you really just compare me to chicken wings?

You say that like it's a bad thing. Chicken wings are the bomb.

— JULIE JAMES

- Whisky
- Whiskey River
- Whiskey Ditch
- White Lightning
- Sangria
- Margarita
- Mojito
- Black & Tan
- Rum & Coke
- Hooch

- Hoot Wine
- Chardonnay
- Merlot
- Moonshine
- Gin & Tonic
- Vodka Tonic
- Colorado Bulldog
- Fuzzy Navel
- Fish Bowl
- Harvey Wallbanger
- Duck Fart
- Monkey Gland
- Fluffy Critter
- Dirty Shirley
- Alabama Slammer
- Rusty Nail
- Lemon Drop
- Coors
- PBR
- Schlitz
- Olympia
- Hamms
- Dr. Pepper / Dr. "Pecker"

CHAPTER 15
Historical

Boys, I many not know much, but I know chicken poop from chicken salad.

— LYNDON B. JOHNSON

- Eleanor Roosevelt
- Franklin D. Roosevelt
- Georg Washington
- John Hancock
- Thomas Jefferson
- John Adams
- Abraham Lincoln
- Betsy Ross
- Hamilton
- Benjamin Franklin
- Paul Revere

- Thomas Edison
- John D. Rockefeller
- Henry Ford
- Susan B. Anthony

CHAPTER 16
Based on Games

Ginger: Listen. We'll either die free chickens or we die trying.

Babs: Are those the only choices?

— CHICKEN RUN

- Badminton
- Lawn Dart
- Parcheesi
- Battleship
- Checkers
- Cribbage
- Cricket
- Tag
- Kick The Can
- Cornhole

- Croquet
- Shuffleboard
- Hopscotch
- Roblox
- Minecraft
- Fortnite
- Grand Theft Auto

CHAPTER 17
Literary Great and Inspiring Artists

I don't know which is more discouraging, literature or chickens.

— E.B. WHITE

- Angie Dickinson
- Harriet Beecher Stowe
- Jane Austen
- Virginia Woolf
- Mary Shelley
- Emily Bronte
- Agatha Christie / "Eggatha" Christie
- JK Rowling
- Harper Lee
- Charlotte Bronte
- Gertrude Stein
- Angela Lansbury

- Emily Dickinson
- Edgar Allen Poe / "Eggdar" Allen Poe
- Picasso
- Vincent van Gogh
- Leonardo da Vinci
- Michelangelo
- Andy Warhol
- Georgia O'Keefe
- Jackson Pollock
- Mona Lisa

CHAPTER 18
Sports Greats

Rocky: Now, the most important thing is, we have to work as a team, which means: you do everything I tell you.

Fetcher: Birds of a feather flop together.

— CHICKEN RUN

- Larry "Bird"
- Magic Johnson
- Michael Jordan
- Dennis Rodman
- Bill Russel
- Kareem Abdul-Jabarr
- Wayne Gretzky
- Phil Mickelson
- Tiger Woods

- Arnold Palmer
- Jerry Rice
- Jim Brown
- Walter Payton
- Barry Sanders
- Joe Greene
- Tom Brady
- Peyton Manning
- Brett Farve
- Johnny Unitas
- Terry Bradshaw
- Joe Montana
- John Elway
- Dan Marino
- Joe Namath
- Brian "The Boz" Bosworth
- Ty Cobb
- Babe Ruth
- Nolan Ryan
- Shoeless Joe Jackson
- Pete Rose
- Ken Griffey Jr.
- Roberto Clemente
- Sandy Koufax
- Rogers Clemens
- Derek Jeter
- Pele
- Flo Jo
- Roger Federer
- Steffi Graf
- Serena Williams

- Billie Jean King
- Martina Navratilova
- Katie Ledecky
- Mary Lou Retton
- Nadia Comaneci
- Dorothy Hamill
- Kristi Yamaguchi
- Tara Lipinski
- Michelle Kwan
- Danica Patrick
- Lisa Leslie

CHAPTER 19
Strong Female Personalities

Nick: Yeah, but you've got to get the chicken first to get the egg, and then you get the egg to get the chicken out of.

Fetcher: Hang on. Let's go over this again?

— CHICKEN RUN

- Frida Kahlo
- Ruth Bader Ginsburg
- Maya Angelou
- Katherine Hepburn
- Oprah
- Nancy Pelosi
- Mary Queen of Scots
- Elizabeth I
- Rosa Parks

- Marie Curie
- Virginia Woolf
- Margaret Thatcher
- Celie Harris Johnson
- Amelia Earhart
- Princess Diana
- Katherine Johnson
- Jane Goodall
- Ruth Handler
- Hillary Clinton
- Toni Morrison
- Ellen DeGeneres
- Gloria Steinem
- Susan B Anthony
- Estee Lauder
- Florence Nightingale
- Helen Keller
- Cleopatra
- Sacajawea
- Mother Teresa
- Betty Crocker
- Julia Child
- Martha Stuart

CHAPTER 20
Female Leads

I'll change you from a rooster to a hen with one shot!

— DOLLY PARTON, 9 TO 5

- Betty White
- Elizabeth Taylor
- Ingrid Bergman
- Tippy Hedren / "Chicky Hen-dren")
- Bette Davis
- Meryl Streep / Meryl "Cheep"
- Anjelica Huston
- Greta Garbo
- Marlene Dietrich
- Jean Harlow
- Grace Kelly
- Judy Garland
- Liza Minelli

- Lauren Bacall
- Hedy Lamarr
- Sophia Loren
- Joan Crawford
- Marilyn Monroe
- Audrey Hepburn
- Lucille Ball
- Carol Burnette
- Gilda Radner
- Wanda Sykes
- Annie Potts
- Darryl Hannah
- Shirley MacLaine
- Sissy Spacek
- Diane Keaton
- Mia Farrow
- Goldie Hawn
- Jamie Lee Curtis
- Jane Fonda
- Lily Tomlin
- Dolly Parton
- Bette Mildler
- Sally Field
- Sigourney Weaver
- Geena Davis
- Whoopi Goldberg
- Pamela Anderson
- Jennifer Aniston / "Henifer" Aniston
- Heather Locklear / "Feather" Locklear
- Gwyneth Paltrow / Gwyneth "Poultry"

CHAPTER 21
Fashion Icons and Luxury Cars

Ace: Are you ready to rock?

Hollywood Runt: Ain't no mountain high enough. Ain't no valley low.

— CHICKEN LITTLE

- Gucci
- Versace
- Dior
- Prada
- Burberry
- Chanel
- Louis Vuitton
- Giorgio Armani
- Calvin Klein
- Valentino

- Hubert de Givenchy
- Vera Wang
- Oscar de la Renta
- Jimmy Choo
- Manolo Blahnik
- Christian Louboutin
- Armani
- Rolex
- Tiffany
- Swarovski

Name your flock after these icons – then, if you're sarcastic) (like me), name "that one chicken":

- Kmart
- Walmart
- Dollar Store
- Salvation Army
- Costco
- Target
- Kohl's
- TJ Maxx
- Old Navy
- Gap
- Forever 21
- Timex

LUXURY VEHICLES

- Porsche

- Ferrari
- Mercedes
- Bentley
- BMW
- Maserati
- Lamborghini
- Cadillac
- Aston Martin
- Lexus
- Jaguar
- Roll-Royce
- Lincoln
- Bugatti
- Audi

Less than Luxury Brands/Models:

- POS
- Kia
- Hyundai
- Honda
- Impala
- Chevy Chevette
- Ford Taurus
- Ford Pinto
- VW Bug

CHAPTER 22
Fictional Characters

> We all like chicken.
>
> — MALCOLM X

- Buckbeak
- Jean-Luc "Peckhard" aka The Captain
- Nancy Drew
- Indiana Jones
- Magnum PI
- Bart Simpson
- Alf
- Mary Poppins / Mary "Poopins"
- Urkel
- Punky Brewster
- Pretty Woman
- Scarlett O'Hara
- Maria Von Trapp

- Jackie Brown
- Erin Brockovich
- Elle Woods
- Jessica Rabbit
- Ariel
- Hermione Granger
- Harry Potter
- Voldemort
- Rasputin
- Katniss Everdeen
- Abominable
- Yeti
- Loch Ness
- Princess Leia / Princess "Lay-a"
- Luke Skywalker
- Hans Solo
- Jabba the Hut
- Baby Yoda
- Mandalorian
- Darth Vader
- Jar Jar Binks
- Lara Croft
- Leeloo
- Mulan
- Sarah Conner
- Terminator
- Black Widow
- Trinity
- Neo
- Aeon Flux
- Ursula

- Bilbo Baggins
- Gandalf / Gandalf the Grey
- Frodo
- Legolas
- Aragorn
- Galadriel
- Smaug

* * *

If you enjoyed this book, please consider leaving a review.

* * *

Bibliography

9 to 5. Directed by Colin Higgins, 20th Century Studios, 19 Dec. 1980.

"100 Best Chicken Quotes, Sayings and Phrases | Kidadl." *Kidadl.com*, kidadl.com/quotes/best-chicken-quotes-sayings-and-phrases.

"A Quote by Lyndon B. Johnson." *Www.goodreads.com*, www.goodreads.com/quotes/85046-i-may-not-know-much-but-i-know-chicken-shit. Accessed 13 June 2023.

Beastie Boys. *5-Piece Chicken Dinner*. 1989.

Buffett, Jimmy. *Life Is Just a Tire Swing*. 1974.

Cash, Johnny. *The Chicken in Black*. 1998.

Chicken Little. Directed by Mark Dindal, Walt Disney Pictures, Walt Disney Studios Motion Pictures, 30 Oct. 2005.

"Chicken Quotes." *BrainyQuote*, www.brainyquote.com/topics/chicken-quotes.

Chicken Run. Directed by Nick Park and Peter Lord, DreamWorks Pictures, Pathé, Universal Pictures, 20th Century Studios, 20th Century Home Entertainment, 8 Dec. 2000.

Dead Kennedys. *Chicken Farm*. 1985.

Dunn, Katherine. *Geek Love*. London, Abacus, 2015.

Finn, Amy. "120 Chicken Quotes to Make You Appreciate Them." *Www.quoteambition.com*, 22 July 2021, www.quoteambition.com/chicken-quotes/.

---. "120 Chicken Quotes to Make You Appreciate Them." *Www.quoteambition.com*, 22 July 2021, www.quoteambition.com/chicken-quotes/.

Fitzgerald, Ella. *Cabin in the Sky*. 1940.

Gucci Mane. *I Move Chickens*. 2007.

"Henny Youngman Quotes." *BrainyQuote*, www.brainyquote.com/quotes/henny_youngman_106841. Accessed 12 June 2023.

James, Julie. *About That Night*. Penguin, 3 Apr. 2012.

"Jean-Michel Basquiat Quotes." *BrainyQuote*, www.brainyquote.com/quotes/jeanmichel_basquiat_543053. Accessed 13 June 2023.

Jordan, Louis . *Ain't Nobody Here but Us Chickens*. 1956.

Little Feat. *Dixie Chicken*. 1973.

BIBLIOGRAPHY

Nas. *Fried Chicken*. 2008.

"Paul Simon Quotes." *BrainyQuote*, www.brainyquote.com/quotes/paul_simon_312941. Accessed 12 June 2023.

The Ozark Mountain Daredevils. *Chicken Train*. 1980.

Thomas, Rufus. *Do the Funky Chicken*. 1970.

X, Malcolm, and Alex Haley. *The Autobiography of Malcolm X*. 1965. New York, Ballantine Books, 2015.

Zac Brown Band. *Chicken Fried*. 2005.

---. *Sic 'Em on a Chicken*. 2006.

Made in the USA
Monee, IL
07 September 2023